THE ARCHITECTURE OF MOLECULES

Linus Pauling and
Roger Hayward

W. H. FREEMAN AND COMPANY | San Francisco and London

CONTENTS

ATOMS AND MOLECULES

We are now living in the atomic age. In order to understand the world, every person needs to have some knowledge of atoms and molecules.

If you know something about atoms and molecules you can understand the accounts of some of the new discoveries that continue to be made by scientists, and can find pleasure in the satisfaction of your intellectual curiosity about the nature of the world. Many scientists find great happiness through the discovery of some fact or the development of some insight into the nature and structure of the world that had previously not been known to anyone. You may share in this happiness through your appreciation of the meaning and significance of the new knowledge.

During the past fifty years scientists, primarily physicists and chemists, have developed many powerful methods of studying atoms and molecules. Among these methods are atomic and molecular spectroscopy (the measurement and interpretation of the wavelength distribution of the light emitted or absorbed by substances), x-ray diffraction of crystals (the determination of the arrangement of atoms in crystals by the study of their scattering of x-rays), electron diffraction of gas molecules (the determination of the arrangement of atoms in gas molecules by the study of the scattering of electrons by the molecules), and the measurement of the magnetic properties of substances. We shall not discuss the experimental techniques and the methods of interpreting them in this book, but shall present straightaway an account of some of the knowledge about molecular architecture that they have provided.

In most molecules and crystals the constituent atoms are arranged in a well-defined way. The arrangement of the atoms in a molecule or crystal is called its molecular structure or crystal structure. For many molecules and crystals the distances between atoms are known with an accuracy of 1 percent or even 0.1 percent. These are average distances; the atoms in molecules and crystals oscillate about their average positions. The amount of oscillation at ordinary temperature is such as to correspond to about 5 percent variation in the distance between the centers of adjacent atoms.

The statements about molecular architecture that are made in the following pages have a firm basis in experiment and observation and are generally accepted by scientists. Some conventions are used in the drawings; their meaning is described in the text.

No standard scale of linear magnification has been used in the drawings. For some small molecules, such as hydrogen (plate 5) and methane (plate 14) the linear magnification is about 800,000,000; for others, such as the halogens (plate 9), it is about 200,000,000; for most of the other drawings the magnification lies within these limits.

The unit of length used in describing molecules and crystals is the Ångström (symbol Å), named in honor of the Swedish physicist Anders Jonas Ångström (1814–1874). One centimeter is 100,000,000 Å, and one inch is about 254,000,000 Å.

About one million substances have been found in nature or made by chemists. Precise structure determinations have been carried out for about ten thousand substances. The structures of only fifty-six are described in the following pages. These examples of molecular architecture have been selected to give you an idea of the great variety of ways in which atoms can interact with one another, and to emphasize the significance of molecular structure to life. Many important structures, such as those of metals and alloys, are not mentioned in this book.

We hope that you will enjoy this introduction to molecular architecture, and that you will be stimulated to learn more about molecular structure and its significance to the world.

THE ARCHITECTURE OF MOLECULES

A CRYSTAL OF GRAPHITE AND ITS STRUCTURE

1

Graphite is a shiny black mineral. It is also called plumbago and black lead. The name black lead came into use because graphite, like the soft metal lead, leaves a gray streak when it is rubbed over paper. It is the principal constituent of the "lead" of lead pencils.

Graphite is sometimes found as well-developed crystals with the form of a hexagonal prism, as shown at the top of the drawing on the facing page. A graphite crystal can be easily cleaved with a razor blade into thin plates.

Graphite is a variety of carbon. Examination with x-rays has shown that the crystal consists of layers of carbon atoms, with the hexagonal structure shown at the bottom of the drawing. Each atom in the layer has three near neighbors, to which it is strongly bonded. The bond length (distance between centers of adjacent atoms) is 1.42 Å, and the layers are 3.4 Å apart. Each layer may be described as a giant planar molecule, and the graphite crystal may be described as a stack of these molecules. The cleavage into hexagonal plates is easily achieved because it involves only the separation of the planar molecules from one another, without breaking any strong interatomic bonds.

ELECTRONS AND ATOMIC NUCLEI

2

The simplest atom is the hydrogen atom. It consists of a nucleus, called the proton, and an electron. The proton is much heavier than the electron; its mass is 1,836 times the electron mass. The proton has one unit of positive electric charge and the electron has one unit of negative electric charge.

Every atom has one nucleus, which has most of the mass of the atom and has a positive electric charge of Z units. Z is called the atomic number. In an electrically neutral atom there are Z electrons in motion about the nucleus.

The structures of atoms of hydrogen ($Z = 1$), oxygen ($Z = 8$), and uranium ($Z = 92$), as they might be revealed by a gamma-ray snapshot, are shown in the drawing. (The nuclei and electrons are, relative to atoms, far smaller than indicated in the drawing; the nuclear diameters are only about a hundred-thousandth of the atomic diameters, and the electron is even smaller.)

Electrons in atoms move around; they do not remain at a constant distance from the nucleus. In the drawing of the hydrogen atom the electron is indicated at about the average distance from the nucleus, 0.80 Å. The shading indicates roughly where the electron is likely to be. The hydrogen atom does not have a definite radius within which the electron remains, but its radius is conventionally taken to be 1.15 Å. All other atoms except helium are larger.

Hydrogen

Oxygen

Uranium

ELECTRON SHELLS

3

A time exposure of atoms reveals that in general the electrons are concentrated into a series of partially overlapping shells. Hydrogen has one electron in its first and only shell (upper left), and helium has two electrons in this shell. All other atoms also have two electrons in the innermost shell. For oxygen these two electrons are indicated close to the nucleus; the oxygen atom also has six electrons in the outer shell.

In the uranium atom the ninety-two electrons are distributed among six shells; in the drawing the two electrons in the innermost shell cannot be distinguished from the nucleus, because their average distance from the nucleus, which in different atoms is approximately inversely proportional to the atomic number Z, is only about 0.01 Å in uranium.

A table giving the names, symbols, and atomic numbers of the elements can be found following plate 57. The symbol of an element is the initial letter of its name or the initial letter and one other letter. For ten elements the symbol is derived from the Latin name: Na for sodium (*natrium*), K for potassium (*kalium*), Fe for iron (*ferrum*), Cu for copper (*cuprum*), Ag for silver (*argentum*), Au for gold (*aurum*), Hg for mercury (*hydrargyrum*), Sn for tin (*stannum*), Pb for lead (*plumbum*), and Sb for antimony (*stibium*). For one element the symbol is derived from the German name: W for tungsten (*wolfram*).

Hydrogen

Oxygen

Uranium

THE PERIODIC TABLE
OF THE ELEMENTS

4

In 1869 the Russian chemist Dmitri I. Mendelyeev (1834–1907) found
that if the elements are arranged in order of their atomic weights they
show a roughly periodic recurrence of similar physical and chemical
properties. Mendelyeev's arrangement is called the periodic table.

A simple periodic table, with 41 elements designated, is given on the
adjacent page, and a complete periodic table following plate 57. These
tables show the elements in the sequence of their atomic numbers, which
is nearly the same as the atomic-weight sequence. Atomic numbers were
reliably assigned to the elements in 1914.

The elements Li, Na, K, Rb, and Cs, indicated by yellow on the facing
page, are very reactive, soft metals, with low melting points. They are
described as a family or group (Group I), and are called the alkali metals.
The adjacent elements, Be, Mg, and so on, are less reactive and harder,
and have higher melting points than the alkali metals. They are called the
alkaline-earth metals (Group II). The family of elements indicated by
green is the halogen family; the halogens are chemically reactive nonme-
tallic substances, which combine with metals to form salts.

The elements He, Ne, Ar, Kr, Xe, and Rn are called the argonons (or
the inert gases, or the noble gases). Their atoms have little tendency to
form chemical bonds. The small chemical reactivity of the argonons is
attributed to the special stability of groups of 2, 10, 18, 36, 54, and 86
electrons about one atomic nucleus.

Some indication of the periodicity in properties of elements in the
sequence of atomic numbers will be evident in the discussion of valence,
covalent radii, and packing radii of atoms in the following pages.

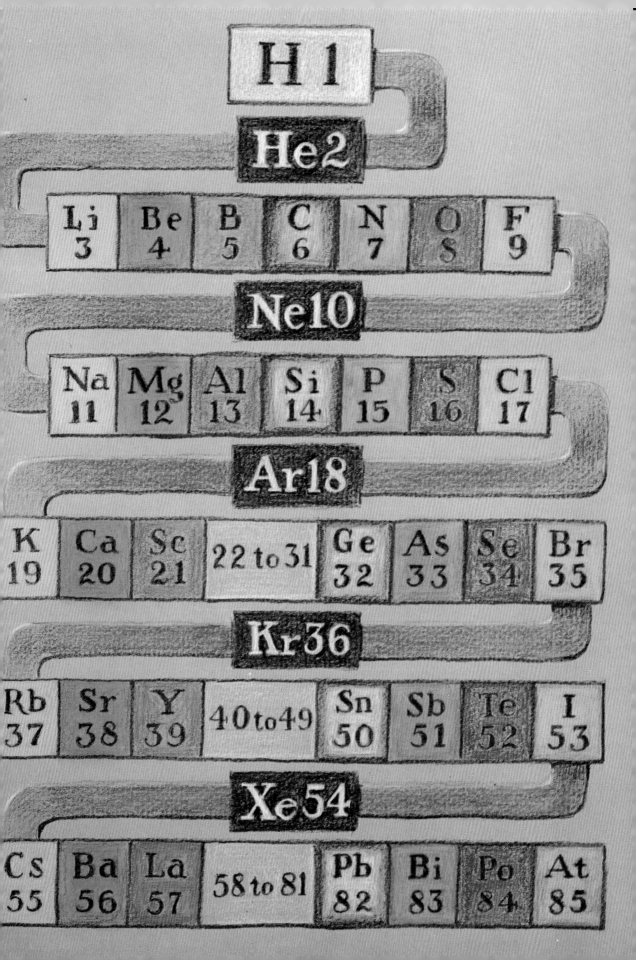

THE HYDROGEN MOLECULE

5

The simplest of all molecules is the hydrogen molecule, which is made of two hydrogen atoms, and is given the chemical formula H_2.

In this molecule there are two protons, 0.74 Å apart, and two electrons, which are held jointly by the two protons and are said to constitute a chemical bond (also called a covalent bond) between them.

In the drawing there are five representations of the structure of the hydrogen molecule. In the first the two H's represent the hydrogen atoms and the dash represents the bond between them. In the second the bond is represented by two dots, which symbolize the two electrons held jointly by the two atoms—they are described as an electron pair shared between the two atoms. In the third, called the ball-and-stick model, the atoms are represented by balls and the bond by a stick. The fourth shows a softened ball-and-stick representation. The fifth, at the bottom of the page, shows the atoms with the effective size that they have in crystalline and liquid hydrogen.

In the crystal and in the liquid, the hydrogen molecules may be described as in contact with one another. The observed distance between the centers of two nonbonded atoms in contact is given reasonably well by the sum of the packing radii of the two atoms. Distances between the centers (the nuclei) of two atoms connected to one another by a chemical bond are given reasonably well by the sum of their covalent-bond radii. (See the tables following plate 57.)

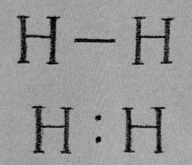

H – H

H : H

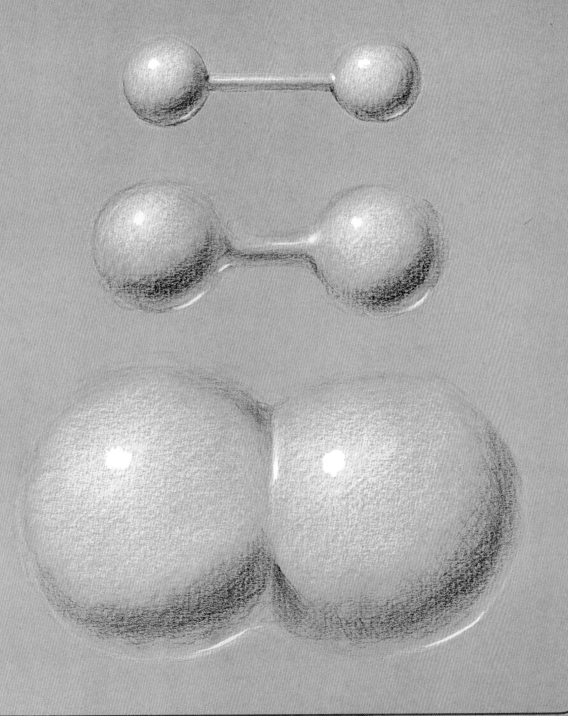

VALENCE

6

Valence is the measure of the number of bonds that an atom can form with other atoms.

The innermost electron shell of an atom can be occupied by only one electron pair, either shared or unshared. Hence the hydrogen atom can form only one bond, using its one electron together with one electron of another atom to form a shared electron pair. Hydrogen has valence one; it is univalent.

Helium, with an unshared electron pair in this shell, has valence zero. The helium atom forms no bonds with other atoms.

The second shell can contain four electron pairs. In neon, $Z = 10$, this second shell is filled with four unshared pairs; hence neon also has valence zero. (Note that, as a conventional simplification, the two electrons of the inner shell are not represented in the drawing for neon and the other atoms with two shells.)

The fluorine atom ($Z = 9$) can form one bond; the atom then has one shared pair and three unshared pairs in its outer shell—it, like hydrogen, is univalent. Similarly, oxygen ($Z = 8$) is bivalent, nitrogen ($Z = 7$) is tervalent (valence three), and carbon ($Z = 6$) is quadrivalent (valence four).

Two sets of atomic symbols are shown on the facing page: the gray symbols, with dots representing electrons, and the red symbols, with dashes representing valence bonds. Both sets are commonly used by chemists, the choice being made as indicated by convenience or habit. Often the two sets are combined, bonds being represented by dashes and unshared electrons by dots.

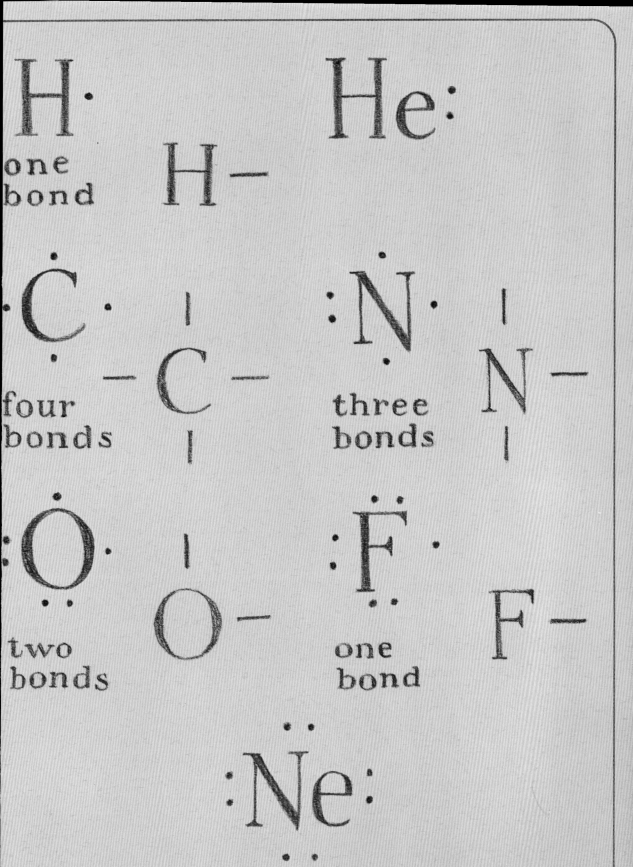

HISTORY OF THE
WATER MOLECULE

7

The ancient philosophers, the alchemists, and the scientists of the seventeenth and eighteenth centuries considered water to be one of the five elements of which the world was assumed to be composed—the other four were variously selected from the group earth, air, fire, ether, acid, iron, mercury, salt, sulfur, and phlogiston.

In 1770 the British scientist Henry Cavendish showed that water is a compound of hydrogen and oxygen. In 1804 another British scientist, John Dalton, assigned to the water molecule the formula shown in the drawing; he assumed the molecule to be made of one atom of hydrogen and one atom of oxygen. By 1860 most chemists had accepted H_2O as the formula of water, and the valence-bond formula H—O—H was being used.

In 1916 the formula with shared-electron-pair bonds was proposed by the American chemist Gilbert Newton Lewis. The large dielectric constant of water was interpreted as showing that the water molecule is not linear, but is bent, with the angle between the two bonds estimated as approximately 110°.

By 1930 the analysis of the absorption spectrum of water vapor had shown that the average value of the bond angle is 104.5° and the average value of the bond length (distance between the nucleus of the oxygen atom and the nucleus of the hydrogen atom) is 0.965 Å.

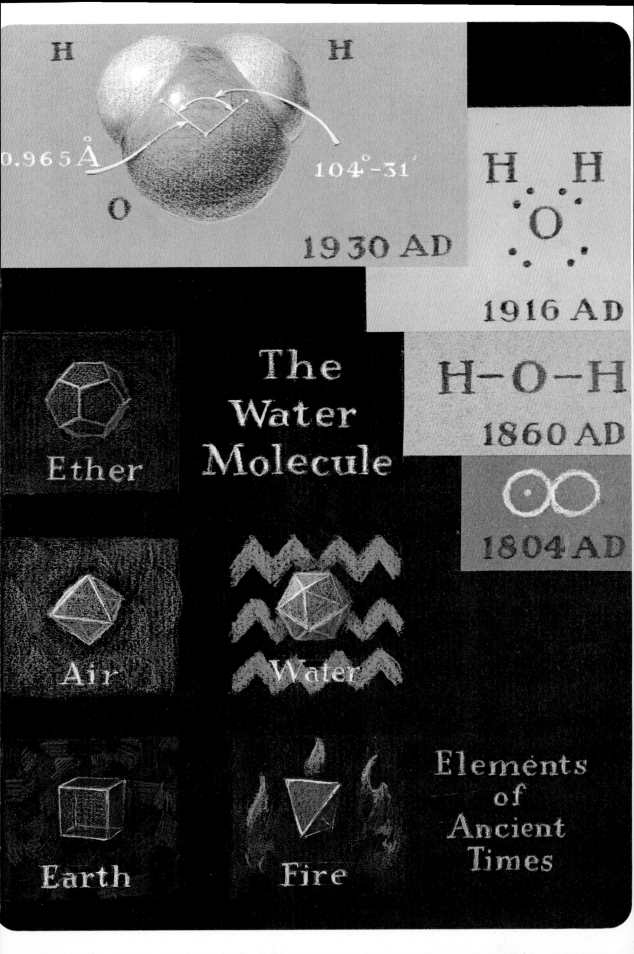

THE BOND ANGLE IN THE WATER MOLECULE AND SIMILAR MOLECULES

8

The observed values of the bond angles in the water molecule and the molecules of hydrogen sulfide, hydrogen selenide, and hydrogen telluride are shown in the drawing. These molecules can be represented by the formula

$$H \diagdown \underset{..}{\overset{}{X}}{..} \diagup H \qquad (X = O, S, Se, \text{ or } Te)$$

The central atom has two unshared electron pairs and two shared pairs in its outer shell.

It is thought that the normal value of the bond angle for an atom with both unshared and shared pairs in its outer shell is 90°, as found in H_2Te and approximated in H_2Se and H_2S.

The large value 104.5° for H_2O may be caused by repulsion between the hydrogen atoms. The distance between the hydrogen atoms in this molecule, 1.53 Å, is much less than the contact distance between hydrogen atoms, 2.3 Å (twice the packing radius, 1.15 Å, of the hydrogen atom). The overlapping of the electron shells of the two hydrogen atoms produces a repulsive force between them that increases the bond angle.

Smaller effects of this sort are seen in H_2S, with hydrogen-hydrogen distance 1.93 Å, and H_2Se, with hydrogen-hydrogen distance 2.10 Å. In H_2Te the hydrogen-hydrogen distance, 2.36 Å, is larger than the contact distance, and the bond angle has the value 90°.

The increase in size of related atoms with increase in atomic number is shown by the bond lengths for these four molecules. The values found by experiment, shown in the adjacent drawing, are very nearly equal to the corresponding sums of covalent radii given in the table following plate 57.

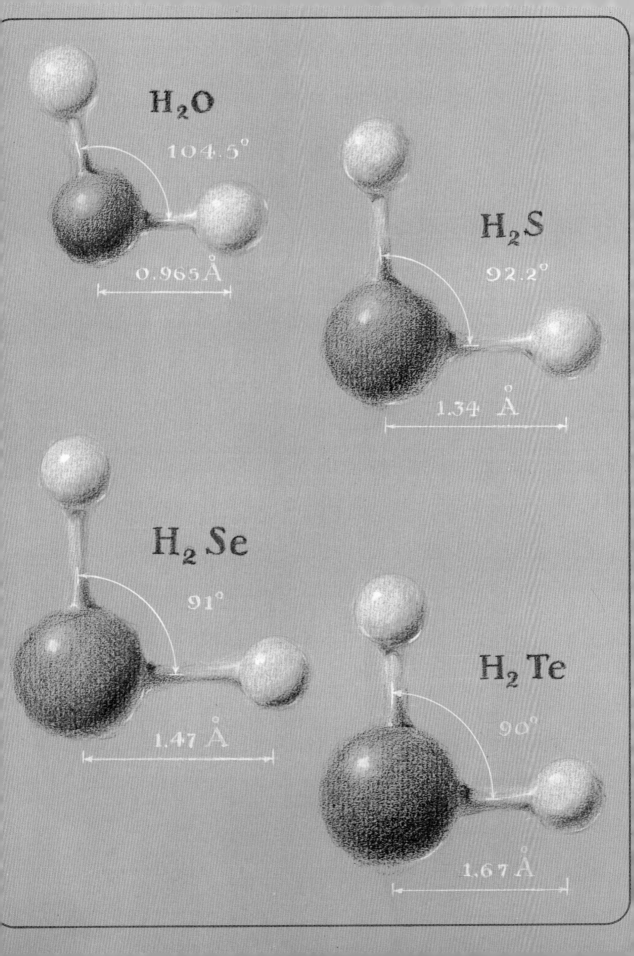

H_2O

104.5°

0.965 Å

H_2S

92.2°

1.34 Å

H_2Se

91°

1.47 Å

H_2Te

90°

1.67 Å

THE HALOGEN MOLECULES

9

The elements fluorine, chlorine, bromine, iodine, and astatine are called the halogens. The word halogen means "salt former," from the Greek words *hals*, salt, and *genes*, producing.

The halogens are univalent. They form the diatomic molecules F_2, Cl_2, Br_2, I_2, and At_2. The first four of these molecules are represented in the drawing.

The molecules have the electronic structure

$$\ddot{:}\overset{..}{X}\!\!-\!\!\overset{..}{X}\ddot{:}$$

The measured bond lengths are 1.43 Å for F_2, 1.99 Å for Cl_2, 2.28 Å for Br_2, and 2.67 Å for I_2. The packing radii, which represent reasonably well the contact distances between the molecules in the halogen crystals, are 1.36 Å for F, 1.81 Å for Cl, 1.95 Å for Br, and 2.16 Å for I.

Astatine is an unstable element, which does not occur in nature. Small amounts of the element have been made, but the molecular properties of At_2 have not yet been determined.

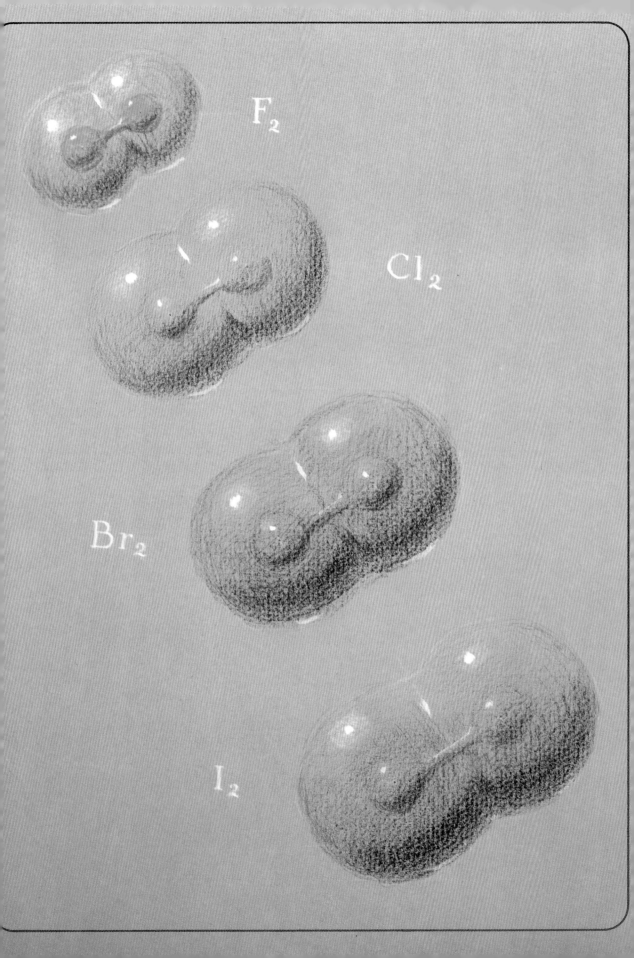

F₂

Cl₂

Br₂

I₂

THE SULFUR MOLECULE

10

The element sulfur is usually seen as yellow crystals or a yellow crystalline powder. Each molecule in these crystals contains eight atoms; its formula is S_8.

These S_8 molecules are present also in solutions of sulfur in solvents such as carbon disulfide and chloroform, and in sulfur vapor at a moderate temperature. Sulfur vapor also contains S_6 molecules and S_2 molecules, in amounts increasing with increase in the temperature.

The S_8 molecule has the form of a staggered ring, as shown in the drawing. The S—S bond length is 2.08 Å. The bond angle, 102°, is larger than the expected value, 90°, because of the crowding of nonbonded sulfur atoms. In the lower drawing of S_8 the heavy lines show the bonds and the light lines indicate pairs of atoms in nonbonding contact. In the upper drawing the atoms are drawn with their packing radii, in order to show the crowding of nonbonded atoms.

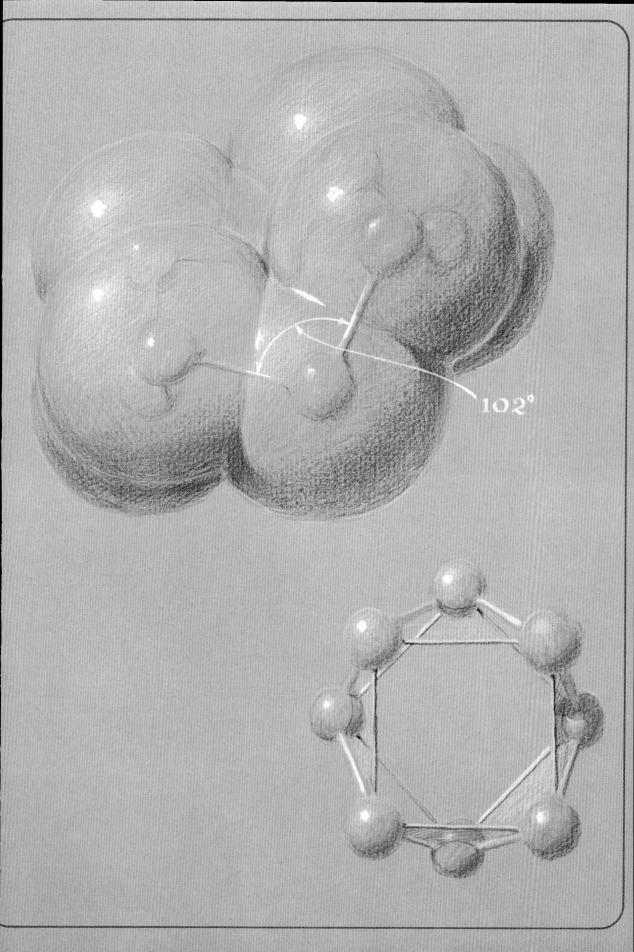

102°

MOLECULES CONTAINING NITROGEN OR PHOSPHORUS ATOMS

11

The nitrogen atom has an electronic structure such as to permit it to form three bonds. Two molecules containing nitrogen atoms are represented in the drawing on the facing page, together with one molecule containing phosphorus atoms.

Ammonia, NH_3, is an important compound of nitrogen. Its most important use is as a fertilizer. It has the electronic structure

$$:N\overset{\displaystyle H}{\underset{\displaystyle H}{-\!H}}$$

in which the nitrogen atom has in its outer shell one unshared electron pair and three shared pairs. The fact that the observed bond angle, 107°, is larger than the expected value, 90°, presumably has the explanation given for the water molecule (plate 8). The values for the related molecules PH_3, AsH_3, and SbH_3 are 93°, 92°, and 91°, respectively. The N—H bond length is 1.00 Å.

Elementary nitrogen is diatomic, with formula N_2. Its structural formula,

$$:N\equiv N:$$

corresponds to the sharing of three electron pairs between the two nitrogen atoms; these two atoms are said to be connected to one another by a triple bond. In the drawing the triple bond is shown as consisting of three bent single bonds. The interatomic distance in N_2 is 1.10 Å.

The element phosphorus when condensed from the vapor forms waxy, colorless crystals, called white phosphorus. These crystals and the vapor contain P_4 molecules, with the tetrahedral structure shown in the drawing. Each of the six interatomic distances in the molecule has the value 2.20 Å.

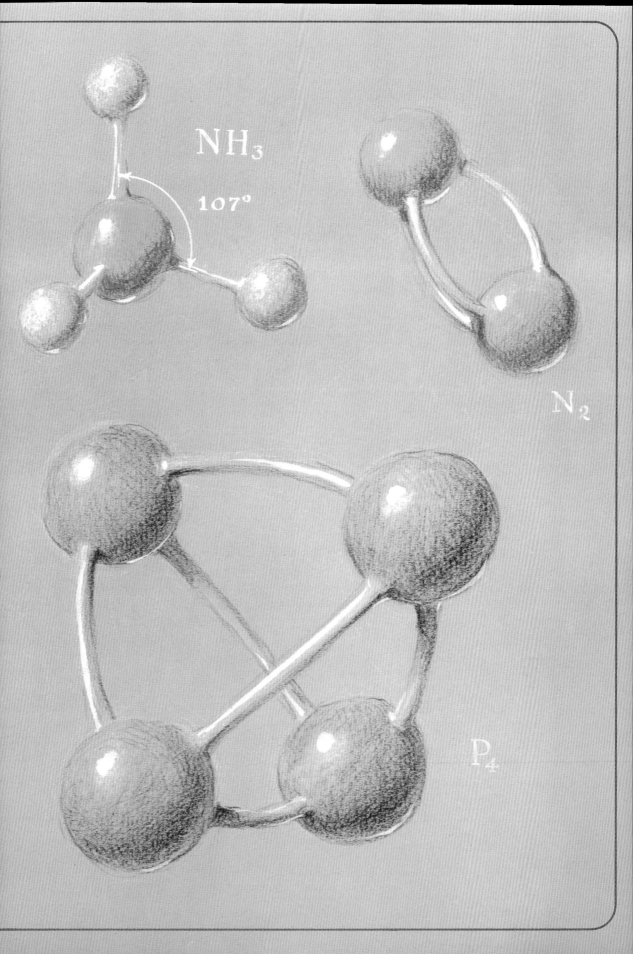

NH$_3$

107°

N$_2$

P$_4$

THE REGULAR POLYHEDRA.
THE CUBE

12 The regular polyhedra are those polyhedra for which all corner angles are equivalent to one another, all edges are equivalent, and all faces are equivalent. The faces are regular polygons. There are five regular polyhedra: the tetrahedron, the cube, the octahedron, the icosahedron, and the pentagonal dodecahedron. There are drawings of them inside the front and back covers of this book. The importance of these polyhedra to molecular architecture is illustrated in later sections of this book.

The Greek philosopher Pythagoras (*circa* 582–500 B.C.) and his students studied the regular polyhedra and introduced them into the Pythagorean cosmology as the symbols of the five elements: the tetrahedron for fire, the cube for earth, the octahedron for air, the icosahedron for water, and the dodecahedron for ether (plate 7). Plato (427–347 B.C.) and the members of his school discussed the regular polyhedra with such vigor as to have caused them to be called the Platonic solids for over 2,300 years.

The cube, which is represented in the adjacent drawing, is the most familiar of the regular polyhedra. It has eight corners, twelve edges, and six faces, which are squares. It has three fourfold axes of rotational symmetry, four threefold axes, six twofold axes, and several symmetry planes. (A figure has an *n*-fold axis of rotational symmetry if rotation through the one-*n*th part of a revolution about an axis produces a figure identical to the original.)

Many crystals have structures that are closely allied to the geometry of the cube. An example will be discussed later (plates 27 and 28).

THE TETRAHEDRON

13

The regular tetrahedron has four corners, six edges, and four faces, which are equilateral triangles. It has four threefold axes of rotational symmetry, and other symmetry elements. This polyhedron has great importance to molecular architecture, as shown by the following example and the discussion on later pages.

One tetrahedral molecule has already been mentioned, the P_4 molecule (plate 11). In this molecule each of the four phosphorus atoms forms three bonds, one with each of the three other phosphorus atoms; the electronic structure is

If the six bonds have the same length they must correspond to the six equivalent edges of the regular tetrahedron, and the phosphorus atoms are thus constrained to occupy the four corners of the tetrahedron.

The relation between the tetrahedron and the cube is shown in the adjacent drawing. This relation permits us to evaluate easily some dimensional properties of the tetrahedron: we note, for example, that the edge of the tetrahedron is $\sqrt{2}a$, where a is the edge of the cube, and the distance from the center of the tetrahedron to a corner is $\sqrt{3}a/2$; hence the ratio of edge to distance from center to corner is $2\sqrt{2}/\sqrt{3}$, which is 1.633.

We may ask why the octatomic molecule P_8, with a cubic structure, is not known to exist and is presumably less stable than the tetrahedral molecule P_4, even though the bending of the bonds in this tetrahedral molecule must involve some strain and some consequent degree of instability. The likely answer is that in the hypothetical cubic molecule P_8 the pairs of atoms related to one another by the diagonal of a square would be only 3.11 Å apart (with the P—P bond length 2.20 Å, as observed in P_4). This distance is so much less than the contact distance 3.8 Å for nonbonded phosphorus atoms (see table of packing radii following plate 57) as to produce a greater strain than the bond-bending strain in P_4.

THE METHANE MOLECULE

14

Methane, which has the formula CH_4, is the simplest hydrocarbon (a hydrocarbon is a compound of hydrogen and carbon). It is a constituent of natural gas and petroleum, and is produced in marshes and stagnant ponds by the decomposition of organic matter. It is the main constituent of intestinal gases.

In 1858 the Scottish chemist A. S. Couper invented valence-bond formulas for chemical compounds and wrote the formula

$$\begin{matrix} H & & H \\ & \diagdown \, C \diagup & \\ H & & H \end{matrix}$$

for methane. Then in 1874 the 22-year-old Dutch chemist Jacobus Hendricus van't Hoff pointed out that some properties of substances could be simply explained by the assumption that the four bonds formed by a carbon atom are directed toward the corners of a tetrahedron, with the carbon atom at its center, as shown in the adjacent drawing. The structure theory of chemistry has been based on the tetrahedral carbon atom ever since.

Spectroscopic studies of methane have shown that the carbon-hydrogen bond length is 1.10 Å. The six H—C—H bond angles have the value 109.5° that is characteristic of the regular tetrahedron.

THE STRUCTURE OF DIAMOND

15

Diamond is an extremely hard crystalline variety of carbon. Its crystals have cubic symmetry, whereas those of graphite, the very soft variety of carbon represented in plate 1, are hexagonal.

The diffraction of x-rays by crystals was discovered in 1912, and in 1913 the first determinations of the atomic arrangement in crystals were made by use of this technique by the British physicists W. H. Bragg and W. L. Bragg (father and son). Their work during this first year included the determination of the structure of diamond, as shown in the adjacent drawing.

In this drawing the cubic unit of structure is represented. The structure of an entire crystal is obtained by repeating this cube in such a way as to fill the volume of the crystal.

You can see that there are eight carbon atoms per unit cube. (Note that only one eighth of each corner atom is in the cube, and one-half of each of the atoms at the centers of the cube faces.)

If the cube is divided into eight smaller cubes, each of the smaller cubes has atoms at four of its corners, and four of the cubes have atoms at their centers.

Each carbon atom is surrounded tetrahedrally by four others. The carbon-carbon bond length is 1.54 Å.

A VIEW OF A DIAMOND CRYSTAL

16

The drawing on the facing page shows a diamond crystal as it might appear to a very small person, with height about equal to the diameter of a carbon atom. (To get this view he would have to change some of the properties of light, as well as those of electrons and atomic nuclei.)

The tunnel down which he is looking extends in the direction of a diagonal of a face of the unit cube shown in the preceding drawing.

The tetrahedral arrangement of the four bonds formed by each carbon atom with its four neighbors is clearly seen in this view of the structure of the crystal. Note that the tetrahedra alternate in their orientations.

The bonds in a diamond crystal bind all of the atoms together into a single molecule. To break the crystal requires breaking many carbon-carbon bonds. We have seen (plate 1) that a graphite crystal can be broken (cleaved) simply by separating the planar molecules from one another, without breaking any bonds. Thus the differing structures of diamond and graphite explain the striking difference in their hardness.

THE ETHANE MOLECULE

17

The structure of the ethane molecule illustrates further the structural principles mentioned in the discussion of the methane molecule and the diamond crystal. The carbon-carbon bond length is 1.54 Å, as in diamond, and the carbon-hydrogen bond length is 1.10 Å, as in methane. The H—C—H and H—C—C bond angles have been found by experiment to have the regular tetrahedral value 109.5° to within the experimental uncertainty, about one-half a degree.

This molecule also illustrates another aspect of molecular structure: the restriction of freedom of rotation about the single bond between the two carbon atoms. Chemists used to think that the two CH_3 groups (methyl groups) could rotate freely relative to one another. Thirty years ago it was discovered that the rotation is restricted. The configuration shown in the drawing, called the staggered configuration (it has the aspect

$$\begin{matrix} & H & \\ H & | & H \\ & \diagup\!\!\!\!\diagdown & \\ H & | & H \\ & H & \end{matrix}$$

along the carbon-carbon axis) is more stable than other configurations.

The configuration with the minimum stability, called the eclipsed configuration

is obtained by rotating one methyl group 60°, relative to the other, from the staggered configuration.

THE NORMAL BUTANE MOLECULE

18

Butane is a hydrocarbon found in natural gas and in petroleum. It is used for domestic heating and lighting.

The structure of the normal butane molecule, n-C_4H_{10}, is shown in the adjacent drawing. (The prefixed letter n in the formula is the abbreviation for normal.) The bond lengths and bond angles have the same values as in ethane. The zigzag chain of carbon atoms and the positions of the hydrogen atoms correspond to the stable (staggered) orientation about each of the four carbon-carbon bonds. The n-butane molecule thus presents us with no surprises.

THE ISOBUTANE MOLECULE

19

Isobutane (also written *iso*-butane) has the same composition as *n*-butane, C_4H_{10}; but isobutane and *n*-butane are different substances, with different properties. For example, isobutane crystals melt at $-145°$ Centigrade, and *n*-butane crystals melt at $-135°$ Centigrade.

The existence of two or more substances with the same composition but different properties and different molecular structure is called isomerism. The substances are called isomers.

Two isomers, such as isobutane and *n*-butane, have the same atoms in their molecules, but the atoms are arranged differently. We have seen on the preceding page that in *n*-butane the four carbon atoms are bonded together to form a zigzag chain. In isobutane there is a branched chain of carbon atoms, as shown in the adjacent drawing.

The structural features of the isobutane molecule are otherwise essentially the same as those of *n*-butane: carbon-carbon bond length 1.54 Å, carbon-hydrogen bond length 1.10 Å, bond angles close to 109.5°, and staggered orientation about the carbon-carbon bonds.

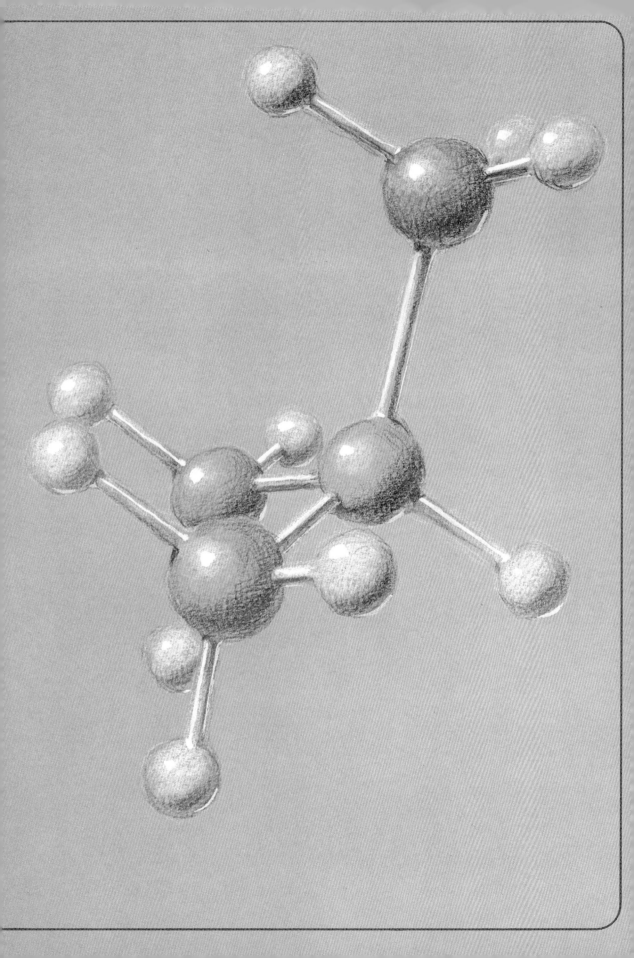

THE CYCLOPROPANE MOLECULE

20

Cyclopropane is a gas at ordinary conditions. It can be condensed to a liquid by increasing the pressure or decreasing the temperature. At one atmosphere pressure it becomes a liquid at −34° Centigrade.

When inhaled it produces unconsciousness (loss of awareness of the environment) and anesthesia (loss of feeling and sensation). It is a good anesthetic agent, except for one disadvantage: its mixtures with air may explode if ignited by an electrostatic spark and thus cause the death of the patient. Accidents of this sort are, however, rare.

The structure of the cyclopropane molecule is shown on the facing page. The molecule contains a ring of three carbon atoms. (The prefix cyclo in the name cyclopropane is from the Greek word *kyklos*, a circle.)

If the bonds of each carbon atom are assumed to be tetrahedral, the carbon-carbon bonds may be described as bent, as shown in the drawing. It is interesting that the measured distance between the carbon atoms in cyclopropane, 1.51 Å, is less than for diamond, ethane, and the two butanes, 1.54 Å. However, the distance measured along the arc of the bent bond is 1.54 Å, so that 1.54 Å may in one sense be considered to be the bond length in the cyclopropane molecule as well as in the other molecules.

A discussion of the stability of cyclopropane in comparison with other cyclic hydrocarbons is given in the text accompanying plate 22.

THE CYCLOPENTANE MOLECULE

21

The cyclopentane molecule contains a ring of five carbon atoms, with the normal carbon-carbon bond length 1.54 Å.

When the structure of this molecule was determined chemists were astonished to discover that the five carbon atoms do not lie in one plane, at the corners of a regular pentagon. The angles at the corners of the regular pentagon have the value 108°, which is so close to the tetrahedral angle 109.5° that very little bending of the bonds would be required for the planar structure, and it had been expected that the molecule would prove to be planar.

Instead, the pentagon was found to be distorted, as shown in the drawing. The distortion places one carbon atom out of the plane of the other four.

The probable explanation of this distortion is that it permits an approximation to the staggered orientation about four of the five carbon-carbon bonds (see Ethane, plate 17). The planar structure would involve the unstable orientation for all five bonds.

Cyclopentane is somewhat less stable than cyclohexane, C_6H_{12}, which has a molecular structure with a zigzag ring of six carbon atoms with no bending of bonds and with the stable (staggered) orientation for each of the six carbon-carbon bonds, as you will see when you turn the page.

THE CYCLOHEXANE
MOLECULE

22

The cyclohexane molecule has the formula C_6H_{12}. The six carbon atoms form a six-membered ring, with the staggered configuration shown in the drawing. This configuration is free from strain: the bond angles all approximate the tetrahedral value, 109.5°, and the orientation of pairs of carbon atoms about each carbon-carbon bond is the staggered, stable one. Cyclohexane is somewhat more stable than cyclopentane, and is considerably more stable than cyclopropane.

There are two reasons why cyclopropane is more unstable than the larger cyclic hydrocarbons: first, the strain of the bent bonds, and second, the unfavorable orientation around them (eclipsed rather than staggered). Hence the energy content of cyclopropane is higher, per unit weight, than that of cyclopentane or cyclohexane. This higher energy content makes cyclopropane a better fuel for rocket propulsion, with liquid oxygen or nitric acid as oxidant, than the other cyclic hydrocarbons.

A LARGE HYDROCARBON RING

23

Large hydrocarbon molecules may assume configurations that are free from strain. An example of a large cyclic hydrocarbon molecule, $C_{24}H_{48}$, is shown in the adjacent drawing. All of the bond angles are tetrahedral, and the orientation about every carbon-carbon bond is the stable one, the staggered configuration found in the ethane molecule and the diamond crystal.

The energy content of this molecule, per unit weight, is nearly the same as that of cyclohexane, which also has a strain-free structure.

In the drawing of $C_{24}H_{48}$ shown on the facing page the atoms have been represented with their packing radii. The molecule is square, and there is a square hole in its center large enough to permit the ring to be threaded by another hydrocarbon molecule. This structural feature leads us to the subject that is discussed on the following page.

A MOLECULE THAT IS TWO LINKED RINGS

24

For one hundred years chemists thought that it might be possible to synthesize a substance with molecules composed of two rings held together only by the geometrical constraint of being linked with one another in the way that two links are held to one another in a chain. Such a substance was made in 1962. The chemists who achieved this goal carried out a chemical reaction involving the formation of a new chemical bond between the two ends of a long molecule, converting it into a ring (the ring containing two oxygen atoms indicated in the adjacent drawing: the formula of the ring is $C_{34}H_{66}O_2$). When this reaction was carried out in a solution containing a large number of cyclic molecules of another kind (the cyclic hydrocarbon $C_{34}H_{68}$), some of the long molecules whose ends were being bonded together were threaded through these ring molecules, so that when the bond was formed the new ring and the old one were linked in the way illustrated in the drawing. This was proved when the chemists who did this work were able to separate from the solution a small amount of a substance with chemical and physical properties that only a linked-ring structure could have.

Substances of this sort are called catenanes (from the Latin word *catena*, a chain).

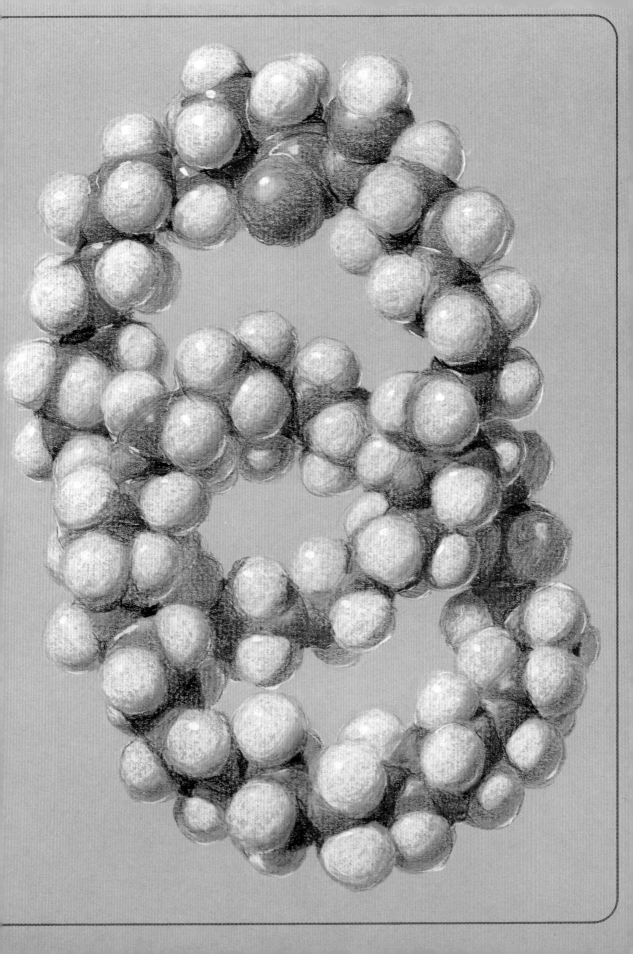

THE POLYOMA VIRUS

25

A polyoma is a tumor or cancer that may grow in various organs of the body, such as the heart and the liver. Recently a substance has been isolated from polyomas of mice and hamsters that seems to be the causative agent of these tumors in both species. It is called the polyoma virus. Much knowledge has been gathered about its molecular structure.

This substance belongs to the class of substances called the nucleic acids. A molecule of a nucleic acid consists of tens or hundreds of thousands of atoms of carbon, hydrogen, oxygen, nitrogen, and phosphorus joined together in a special way.

The units of heredity, called genes, are molecules of nucleic acid. These molecules control the development and growth of living organisms. They cause children to resemble their parents.

The polyoma virus may be a gene that has become abnormal. The molecule of the virus differs from the molecules of the normal gene nucleic acid in that the virus molecules contain rings, whereas the normal molecules are thought to be long chains.

Each virus molecule consists of two rings intertwined in the complex way shown in the drawing. The molecule may be called a catenane, but it is far more complicated than the simple catenane shown in the preceding drawing. Each ring contains about 150,000 atoms.

No one knows why these strange molecules produce cancer in mice and hamsters. Perhaps one of the young readers of this book will discover the mechanism of their cancerogenic action.

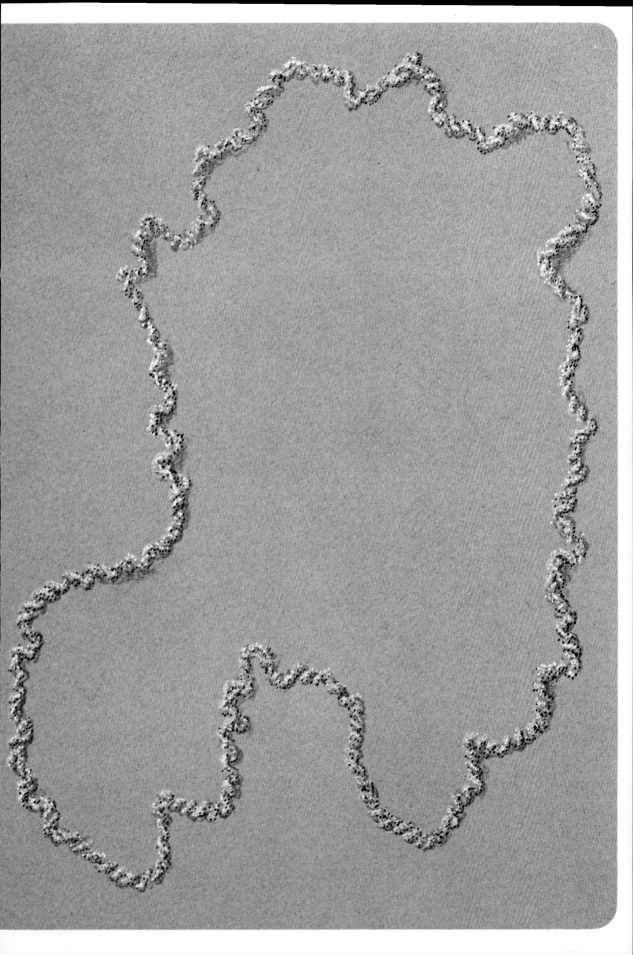

DOUBLE BONDS AND
TRIPLE BONDS

26

The nitrogen molecule, N_2, was described in the discussion accompanying plate 11 as having the structure

$$:N\equiv N:$$

with three bent bonds between the two atoms. Many other molecules contain triple bonds, resembling the bond in the nitrogen molecule, and many contain double bonds.

The hydrocarbon ethylene, C_2H_4, is the simplest substance with a carbon-carbon double bond. It is a gas with the unusual property of causing fruit, such as bananas and oranges, to ripen. Its structural formula is

$$\begin{matrix} H & & H \\ \diagdown & & \diagup \\ & C{=}C & \\ \diagup & & \diagdown \\ H & & H \end{matrix}$$

and its molecular structure is shown in the adjacent drawing.

Acetylene, C_2H_2, is the simplest substance with a carbon-carbon triple bond. It has the structural formula

$$H{-}C\equiv C{-}H$$

and the linear structure shown in the drawing.

The spatial configurations of ethylene (planar) and acetylene (linear) correspond to the theory of the tetrahedral carbon atom; they may be described as two tetrahedra sharing an edge (ethylene) or a face (acetylene).

It is interesting that although the observed carbon-carbon distance in ethylene is 1.33 Å, and that in acetylene is 1.20 Å, the distance measured along the arc of the bent bonds in both has the normal value of the single bond length, 1.54 Å.

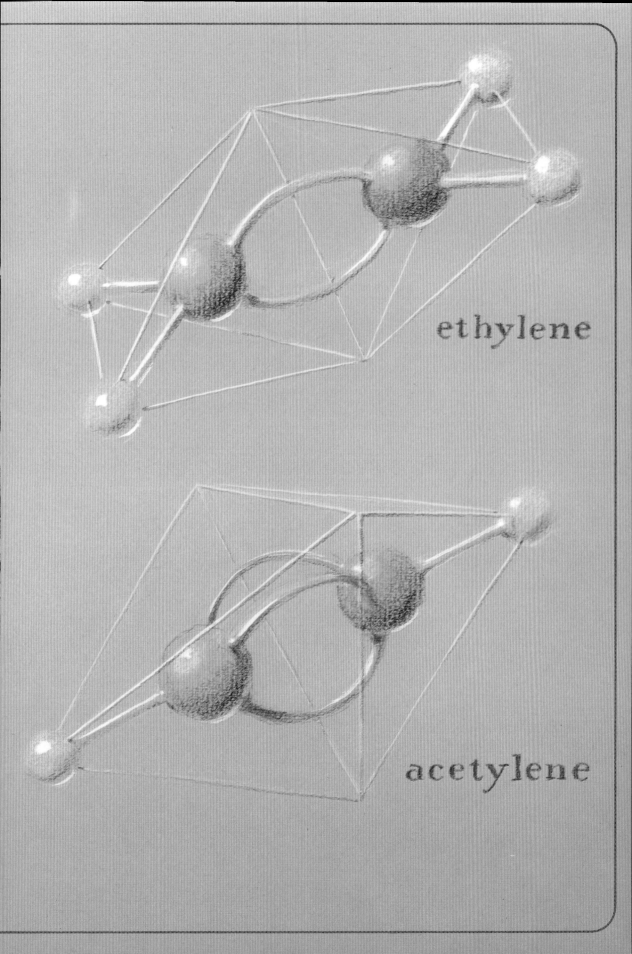

ethylene

acetylene

THE FRAMEWORK OF THE PRUSSIAN BLUE CRYSTAL

27

When the pale violet substance ferric nitrate, $Fe(NO_3)_3 \cdot 6H_2O$, and the yellow substance potassium ferrocyanide, $K_4Fe(CN)_6 \cdot 4H_2O$, are dissolved in water and the solutions are mixed, a precipitate with a brilliant blue color is formed. This precipitated substance is called Prussian blue. It is used as a pigment. Its formula is $KFe_2(CN)_6 \cdot H_2O$.

The x-ray study of the substance has shown that it consists of cubic crystals with the structural framework shown in the adjacent drawing.

This framework illustrates the significance of the cube in molecular architecture. The whole crystal can be described as a three-dimensional cubic lattice. At each corner of each small cube there is an iron atom, and a cyanide group lies along each cube edge.

The bonds in this framework extend along the cube edges. One cube edge can be represented by the formula

$$Fe—C≡N—Fe$$

The observed iron-carbon bond length is 2.00 Å, and the carbon-nitrogen distance is 1.16 Å. This is the value found by experiment for the carbon-nitrogen distance in molecules to which chemists assign a structure with a carbon-nitrogen triple bond, such as the hydrogen cyanide molecule, with formula

$$H—C≡N:$$

(For simplicity the three bent bonds of the triple bond are represented in the drawing by a single bond.)

THE PRUSSIAN BLUE CRYSTAL

28

The crystals of Prussian blue contain potassium ions and water molecules, in addition to the iron-carbon-nitrogen framework.

An ion is an atom or group of atoms with a positive or negative electric charge. An ion is formed from an atom or molecule by removing or adding one or more electrons. A potassium ion, K^+, is a potassium atom from which one electron has been removed, leaving the atom with a positive electric charge. (In Prussian blue the electron has been transferred from the potassium atom to the iron-carbon-nitrogen framework.) In the crystal there is a potassium ion at the center of half of the small cubes. Some of them are illustrated in the drawing.

The other cubical chambers in the crystal are occupied by water molecules. These molecules are not bonded to the other atoms of the crystal. They are present because the crystals were precipitated from an aqueous solution, and some of the water molecules of the solution were entrapped as the framework of the crystal was formed. When a crystal of Prussian blue is heated the water molecules escape from it, leaving the framework with half of its cubical chambers empty.

A crystal of this sort, containing molecules that are not bonded to the framework of the crystal, is called a clathrate crystal (from the Latin word *clathri*, lattice).

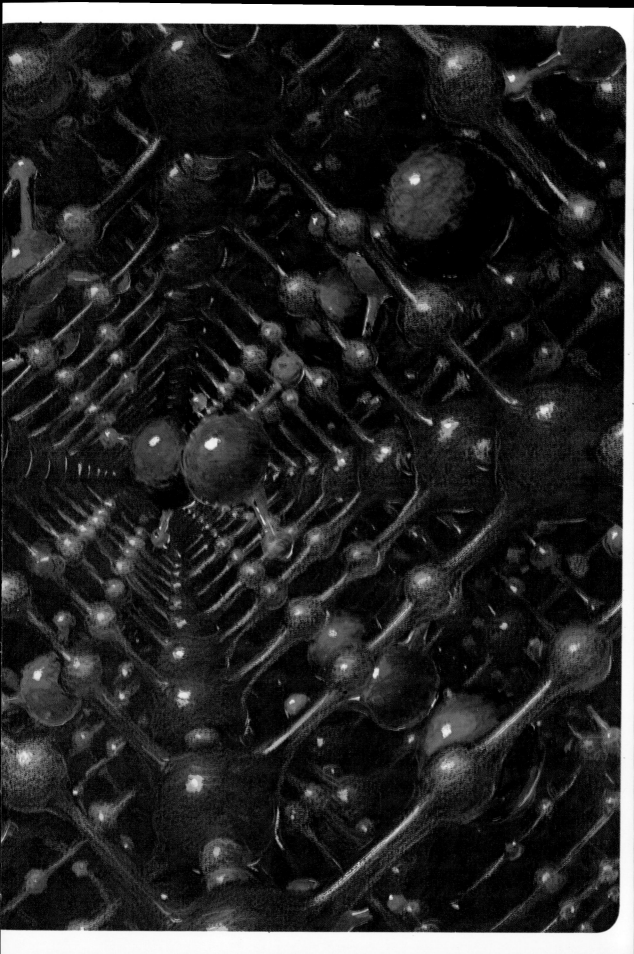

THE OCTAHEDRON

29

The octahedron is one of the regular polyhedra. It has eight equilateral triangular faces, twelve edges, and six corners.

Many molecules are known in which a central atom forms bonds with six other atoms that are arranged about it at the corners of a regular octahedron. An example is sulfur hexafluoride, SF_6. In this octahedral molecule the sulfur atom uses the six electrons of its outer shell to form six bonds, one with each of the six fluorine atoms. This molecule is described as involving octahedral coordination of the six fluorine atoms about the sulfur atom or octahedral ligation of the six fluorine atoms to the sulfur atom.

The number of atoms arranged about a central atom is called the coordination number or the ligancy of the central atom. Octahedral coordination (octahedral ligation) corresponds to ligancy six.

We have already discussed the structure of a crystal containing atoms with octahedral coordination, the Prussian blue crystal. Each of the iron atoms in the framework of this crystal forms six bonds directed toward the corners of a regular octahedron (see plates 27 and 28).

ISOMERIC MOLECULES WITH OCTAHEDRAL COORDINATION

30

Octahedral coordination was discovered by the Swiss chemist Alfred Werner in 1900. He showed that the properties of many substances with complex formulas could be accounted for by assigning them structures involving octahedral coordination. For example, two substances with the molecular formula $Pt(NH_3)_2Cl_4$ were known, an orange substance and a lemon-yellow substance. Werner showed that the difference in their properties could be explained by assigning to each of them an octahedral structure with the platinum atom at the center of the octahedron; he assumed, however, that for one substance the two nitrogen atoms of the ammonia molecules lie at two octahedron corners that define an edge of the octahedron, and that for the other the two nitrogen atoms lie at opposite corners of the octahedron, as shown in the adjacent drawing.

From the properties of the substances Werner concluded that the orange substance is the *cis* isomer (with the two nitrogen atoms defining an edge) and the lemon-yellow substance is the *trans* isomer. (The Latin words *cis* and *trans* mean "on this side" and "on the other side," respectively.) During recent decades the octahedral structures postulated by Werner have been verified by the x-ray diffraction of crystals, and his identification of the *cis* and *trans* isomers has been shown to be correct.

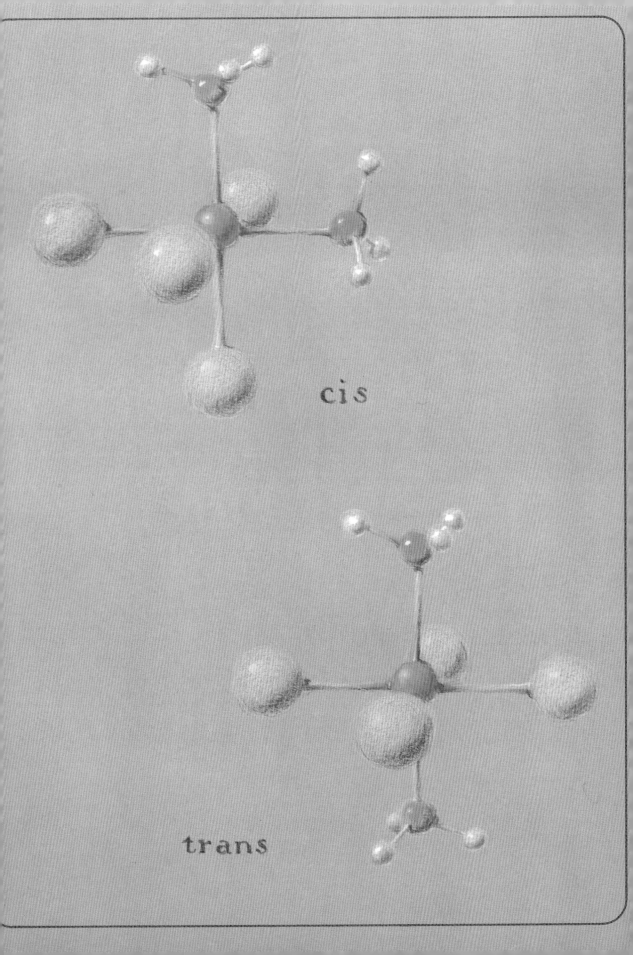

cis

trans

THE HEXAMETHYLENE-TETRAMINE MOLECULE

31

The first determination of the molecular architecture of a compound of carbon was made in 1923, when the structure of the cubic crystal hexamethylenetetramine, $C_6H_{12}N_4$, was determined by x-ray diffraction.

The crystal was found to contain molecules with the structure shown in the adjacent drawing. All bond angles have values close to the tetrahedral value, 109.5°. The carbon-nitrogen bond length, 1.47 Å, and the carbon-hydrogen bond length, 1.10 Å, are those assigned to single bonds between these pairs of atoms (see the table of covalent radii, following plate 57).

The molecule has the symmetry of the regular tetrahedron—four threefold rotation axes, three twofold rotation axes, and some symmetry planes. The four nitrogen atoms lie at the corners of a regular tetrahedron and the six carbon atoms lie at the corners of a regular octahedron. The polyhedron defined by the twelve hydrogen atoms is the truncated tetrahedron, a tetrahedron with each corner cut off by a plane parallel to the opposite face.

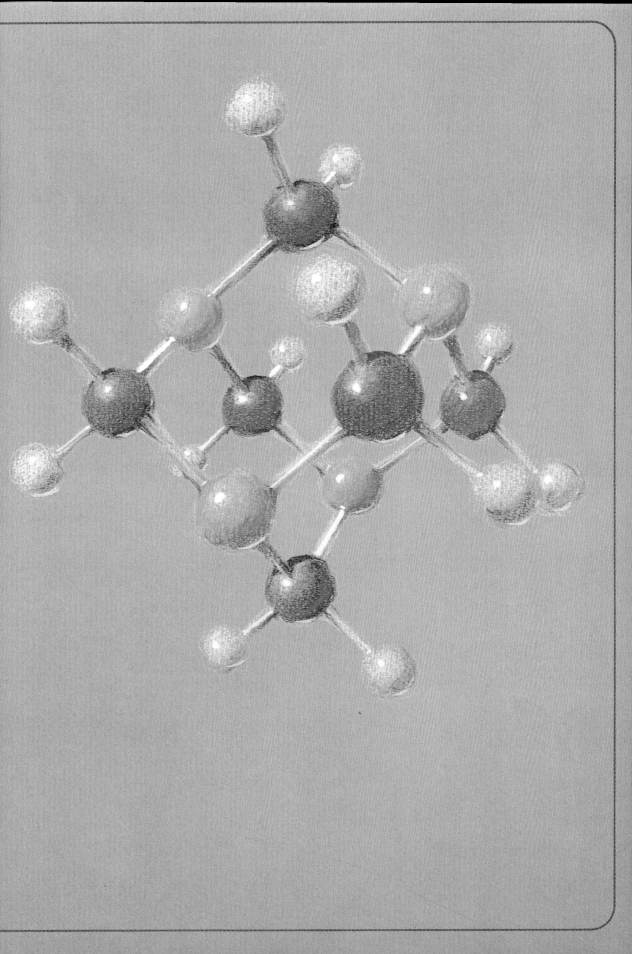

THE ICOSAHEDRON

32

The regular icosahedron, the fourth of the regular polyhedra, has twenty equilateral triangular faces, thirty edges, and twelve corners. Its name is derived from the Greek *eikosi*, twenty, and *hedra*, seat or base. It has many symmetry elements, including six fivefold axes of rotational symmetry, ten threefold axes, and fifteen twofold axes.

The significance of the icosahedron to molecular architecture is illustrated by the discussion of the tetragonal boron crystal, the dodecaborohydride ion, the decaborane molecule, and the tetraborane molecule on the following pages.

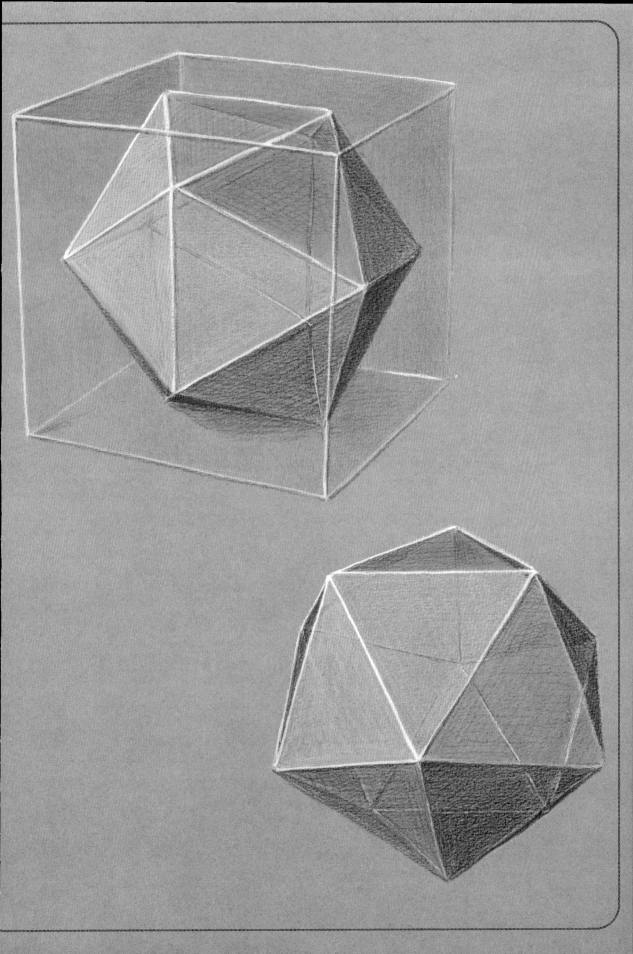

THE TETRAGONAL
BORON CRYSTAL

33

The element boron (atomic number 5) forms crystals that are nearly as hard as diamond. One of the several crystalline modifications has tetragonal symmetry. Its unit of structure is a square prism containing fifty boron atoms, which are arranged as shown in the adjacent drawing.

The fifty boron atoms in the unit comprise four B_{12} groups and two other boron atoms. Each of the B_{12} groups is an icosahedron.

Each boron atom of a B_{12} group is bonded to five neighboring boron atoms in the same group and also to a sixth atom, either an atom of an adjacent B_{12} group or one of the two other atoms. These two atoms that are not in B_{12} groups show tetrahedral ligation.

The boron-boron bonds in this crystal are not ordinary bonds. Boron, with $Z = 5$ and two electrons in its inner shell, has only three valence electrons, and could form only three shared-electron-pair bonds. But most of the atoms form six bonds. We conclude that these bonds are not ordinary covalent bonds, but are bonds of another kind, involving only one electron per bond. They are called half-bonds; a half-bond is only about one-half as strong as an ordinary covalent bond, which involves two electrons (rather than one) shared between two atoms.

The bond length for a normal boron-boron bond estimated from the observed values for other single bonds is 1.62 Å. The larger value 1.80 Å found in the tetragonal boron crystal is reasonable for a half-bond.

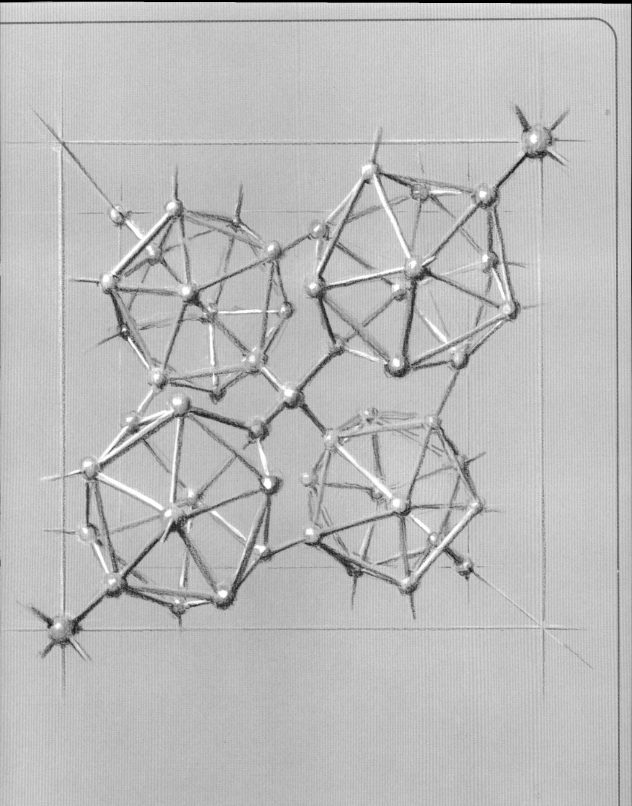

AN ICOSAHEDRAL BOROHYDRIDE ION

34

Among the many interesting compounds of boron is the colorless crystalline substance dipotassium dodecaborohydride, $K_2B_{12}H_{12}$. The x-ray investigation of this crystal has shown that it contains $B_{12}H_{12}^{--}$ ions and potassium ions, K^+. Each $B_{12}H_{12}^{--}$ ion has two extra electrons, which have been transferred to it from two potassium atoms.

The $B_{12}H_{12}^{--}$ ions have the icosahedral structure shown in the drawing. Each boron atom has ligancy six. The boron-boron bonds seem to be half-bonds, as in tetragonal boron; their length is 1.80 Å. The boron-hydrogen bonds have a length of about 1.20 Å.

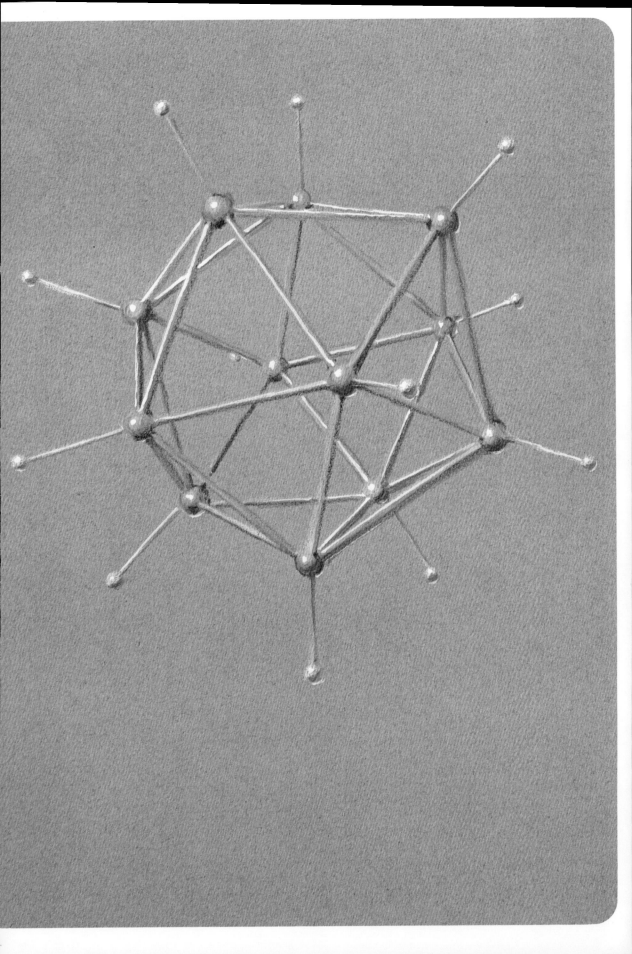

THE DECABORANE MOLECULE

35

The compounds of boron and hydrogen are called the boranes. Many boranes are known, and, although chemists have studied them for many years, they continue to provide puzzling problems.

Decaborane, $B_{10}H_{14}$, has the structure shown in the adjacent drawing. The boron atoms lie at ten of the twelve corners of an icosahedron. The boron-boron distance is 1.80 Å. A hydrogen atom lies 1.18 Å out from each boron atom. Each of the other four hydrogen atoms, called bridging hydrogen atoms, is bonded to two boron neighbors (at a distance of about 1.35 Å), rather than to one. These bonds and the boron-boron bonds seem to be half-bonds.

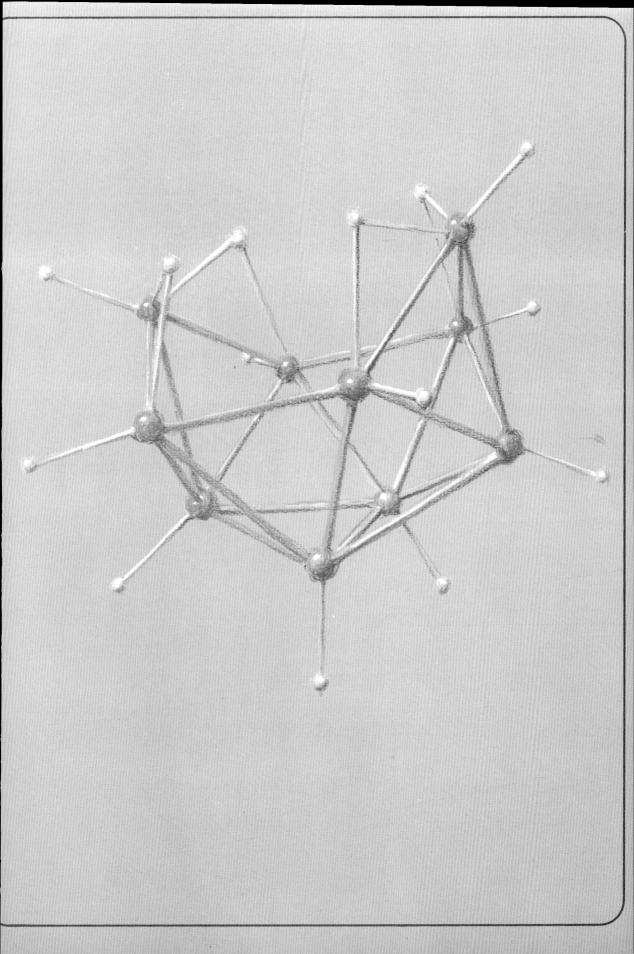

THE TETRABORANE MOLECULE

36

The molecular structure of tetraborane, B_4H_{10}, is shown in the adjacent drawing. The four boron atoms, the four bridging hydrogen atoms, and two of the nonbridging hydrogen atoms lie approximately at ten of the corners of an icosahedron, and the other four hydrogen atoms lie radially out from the boron atoms. The dimensions are very nearly the same as in decaborane.

The boron atoms have ligancy six, as in decaborane and the icosahedral borohydride ion. Ligancy six is found for boron atoms in most other boranes, but ligancy five is found in a few.

The boranes react vigorously with liquid oxygen and other oxidants to form boric oxide, B_2O_3, and water. They have been extensively investigated for possible use as rocket fuels.

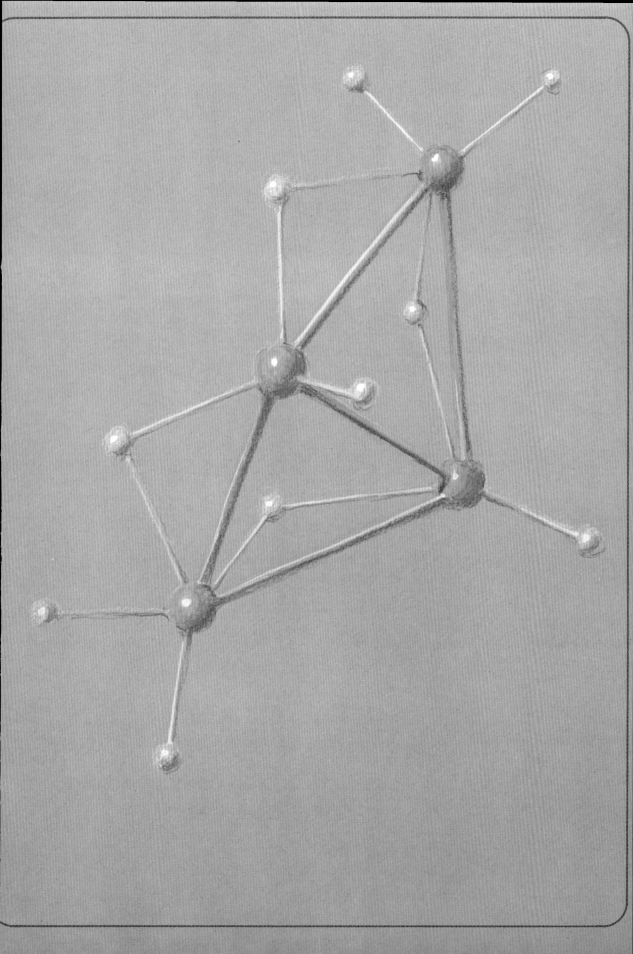

THE CARBON ATOM WITH LIGANCY SIX

37

After sixty-five years of successful use of the theory of the tetrahedral carbon atom, chemists were astonished in 1939 by the report that two American chemists, R. E. Rundle and J. H. Sturdivant, had discovered carbon atoms with ligancy six in a molecule.

Rundle and Sturdivant carried out the determination by x-ray diffraction of the crystal structure of the substance to which the name platinum tetramethyl and the formula $Pt(CH_3)_4$ had been assigned. They found, however, that each molecule contains four platinum atoms and sixteen methyl groups, as shown in the adjacent drawing. The molecular formula is $Pt_4(CH_3)_{16}$.

Twelve of the methyl groups are normal, with tetrahedral carbon atoms, each forming three C—H bonds and one C—Pt bond. The platinum atoms show octahedral coordination. Each of the other four methyl groups forms a bridge connecting three of the platinum atoms to one another. The carbon atom of a bridging methyl group forms six bonds, three with its hydrogen atoms and three with platinum atoms.

The platinum-carbon bond length for the bridging carbon atoms is about 0.2 Å greater than that for the other carbon atoms, indicating that the bridging carbon atoms form fractional bonds (perhaps one-third bonds) with the platinum atoms.

THE FERROCENE MOLECULE

38

The substance ferrocene was first made in 1951. Its molecules have the formula $Fe(C_5H_5)_2$ and the structure shown in the adjacent drawing. The molecule is called a sandwich molecule—it may be described as an atom of iron sandwiched between two hydrocarbon rings.

Many sandwich molecules have been studied during the past decade. However, there is still uncertainty about the nature of the bonds in ferrocene and related substances.

We might think that the bonds are all single bonds. Each carbon atom would then form four bonds, with considerable distortion from the regular tetrahedral directions. There are, however, theoretical arguments against the formation of ten single bonds by an iron atom. Also, the observed iron-carbon bond length, 2.05 Å, indicates that the bonds are approximately half-bonds. The carbon-carbon distance in the rings, 1.44 Å, is intermediate between the value for a single bond, 1.54 Å, and the value for a double bond, 1.33 Å.

THE HYDROGEN BOND

39

An important property of the hydrogen atom was discovered in 1920 by two American chemists, W. M. Latimer and W. H. Rodebush. They found that many properties of substances can be easily explained by the assumption that the hydrogen atom, which is normally univalent, can sometimes assume ligancy two, and form a bridge between two atoms. This bridge is called the hydrogen bond.

The most important hydrogen bonds are those in which a hydrogen atom joins pairs of fluorine, oxygen, or nitrogen atoms together.

Hydrogen fluoride gas has been found to contain not only molecules HF, but also polymers, especially $(HF)_5$ and $(HF)_6$. The structure of $(HF)_5$ is shown in the adjacent drawing. Each hydrogen atom is strongly bonded to one fluorine atom (bond length 1.00 Å) and less strongly to another (bond length 1.50 Å). The bond angle at the hydrogen atom forming a hydrogen bond is usually about 180°.

The hydrogen difluoride ion, FHF^-, exists in crystals (such as KHF_2, which contains the ions K^+ and FHF^-) and in aqueous solution. It has an unusual structure, in that the hydrogen atom (proton) is midway between the two fluorine atoms (fluoride ions). The two fluorine-hydrogen bonds in FHF^- have the length 1.13 Å.

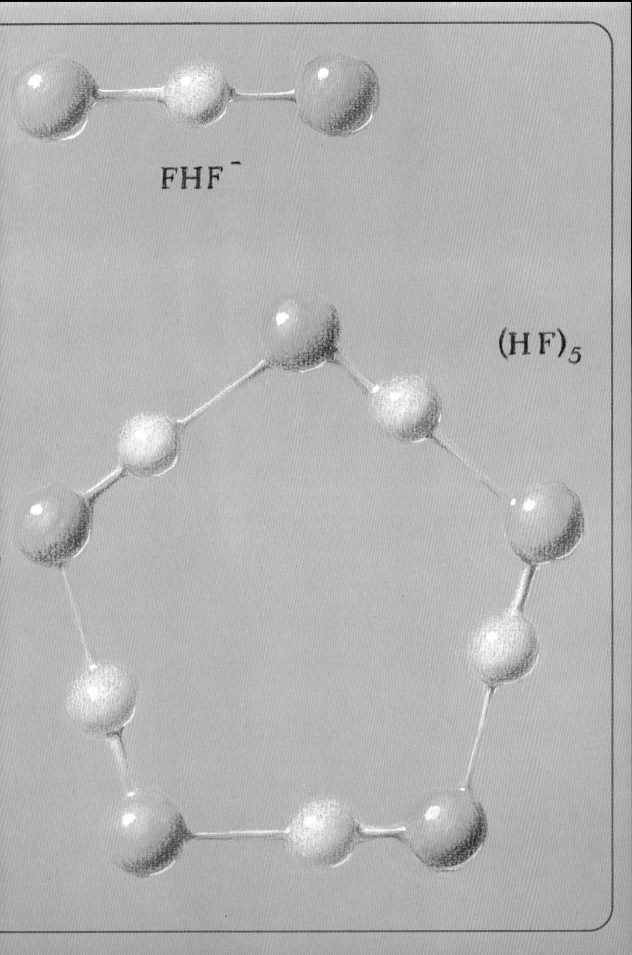

FHF⁻

(HF)₅

THE DOUBLE MOLECULES OF ACETIC ACID

40

Acetic acid is the acid in vinegar. After many years of studying its properties and reactions, chemists were convinced that its molecules could be assigned the structural formula

$$H_3C-C\overset{\displaystyle O}{\underset{\displaystyle O-H}{<}}$$

There were, however, some facts that could not be easily explained by this formula. The density of acetic acid vapor and the properties of solutions of the acid in some solvents indicated that some of the molecules have the formula $C_4H_8O_4$ rather than $C_2H_4O_2$.

The explanation was provided by the theory of the hydrogen bond. According to this theory, two acetic acid molecules with the generally accepted structural formula could combine by forming two hydrogen bonds with one another, to produce a double molecule (called acetic acid dimer), as shown in the adjacent drawing.

The structure of acetic acid dimer and other acid dimers has been verified by the electron diffraction and x-ray diffraction techniques. Each hydrogen atom is 1.00 Å from one oxygen atom and about 1.60 Å from another.

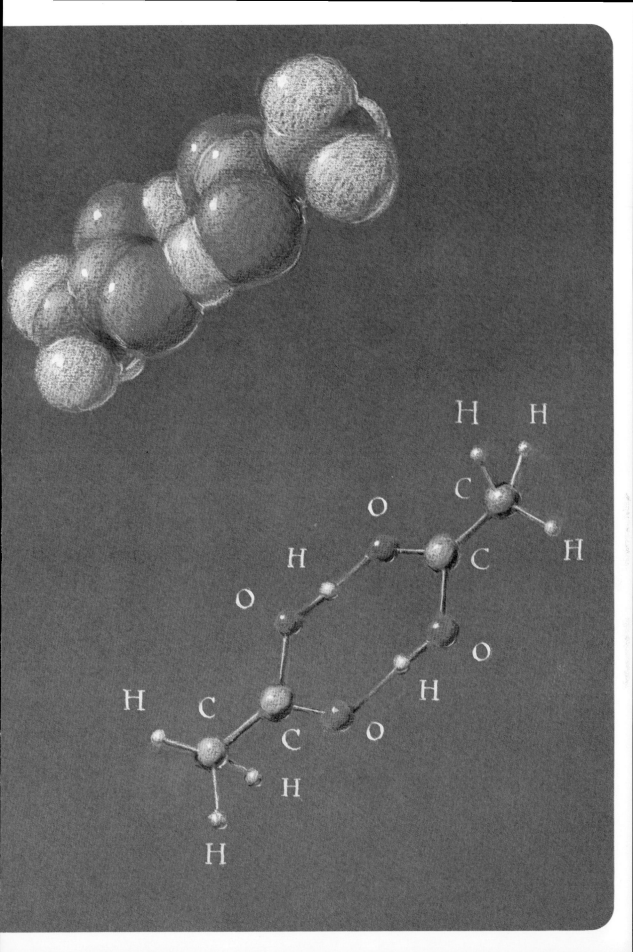

ICE

41

Ice has an interesting and important property that differentiates it from almost all other crystalline substances: ice floats in water. It is less dense than the liquid that it forms on melting, whereas most crystals are denser than their liquids.

The explanation of this property is provided by the crystal structure of ice. The crystal is hexagonal, with the rather open framework structure shown in the adjacent drawing.

Each water molecule in ice is tetrahedrally surrounded by four other water molecules, and connected to them by hydrogen bonds. In each hydrogen bond the hydrogen atom is 1.00 Å from one oxygen atom and 1.76 Å from another.

When ice melts, some of the hydrogen bonds are broken, and in consequence the water molecules are able to pack together more compactly in liquid water (in which the average ligancy of a water molecule is about five) than in ice (ligancy four).

There is an interesting sort of structural disorder in the ice crystal. For each hydrogen bond there are two positions for the hydrogen atom: O—H – – – O and O – – – H—O. If there were no restriction on this disorder there would be 4^N ways of arranging the hydrogen atoms in an ice crystal containing N water molecules ($2N$ hydrogen atoms). But there is a restriction: there must be two hydrogen atoms near each oxygen atom. In consequence there are only $(3/2)^N$ ways of arranging the $2N$ hydrogen atoms in the crystal.

Some of the properties of ice are affected by the disorder of the hydrogen bonds. Measurement of these properties has verified the calculated number of arrangements to within about 1 percent.

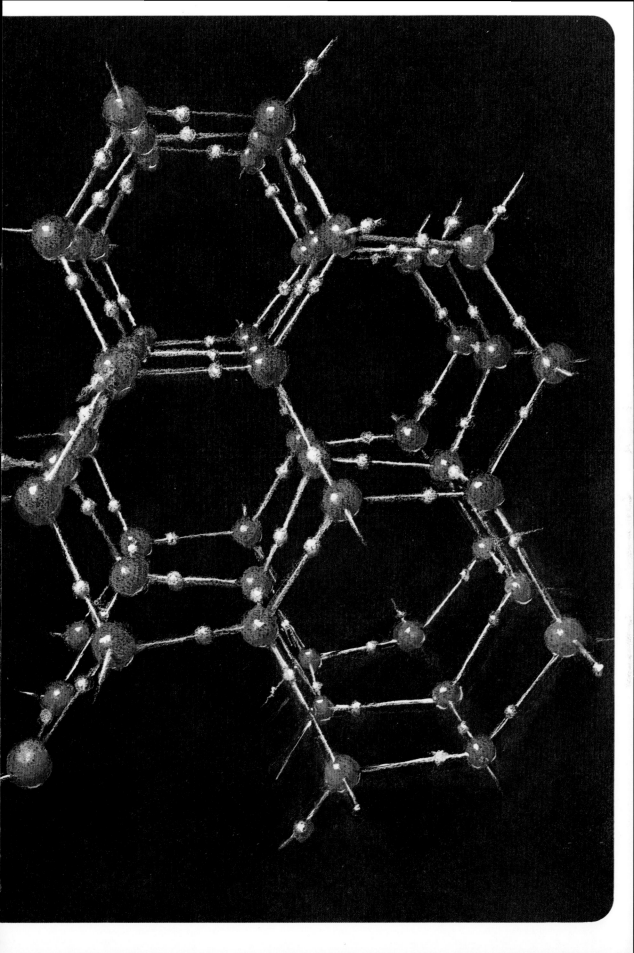

A DENSE FORM OF ICE

42

Several different crystalline modifications of ice are formed under high pressure. Ice II, a kind of ice that forms at about 3,000 atmospheres pressure, has the structure shown in the adjacent drawing.

In this form of ice there are columns of six-molecule hydrogen-bonded rings, rather similar to those in ice I. The columns are pushed more closely together than in ice I. Each water molecule forms hydrogen bonds with four neighboring water molecules, which are at the corners of a distorted tetrahedron. The distortion permits a fifth water molecule to approach rather closely (3.24 Å).

Ice II is about 15 percent denser than liquid water at the same pressure, whereas ordinary ice is 8 percent less dense than water.

There is no randomness in the hydrogen atom positions of ice II. Several other high-pressure forms of ice are known, ice III, ice IV, ice V, ice VI, and ice VII. Their properties indicate that they all have about the same amount of hydrogen-atom randomness as ice I, and that ice II is the only form of ice with completely ordered hydrogen bonds.

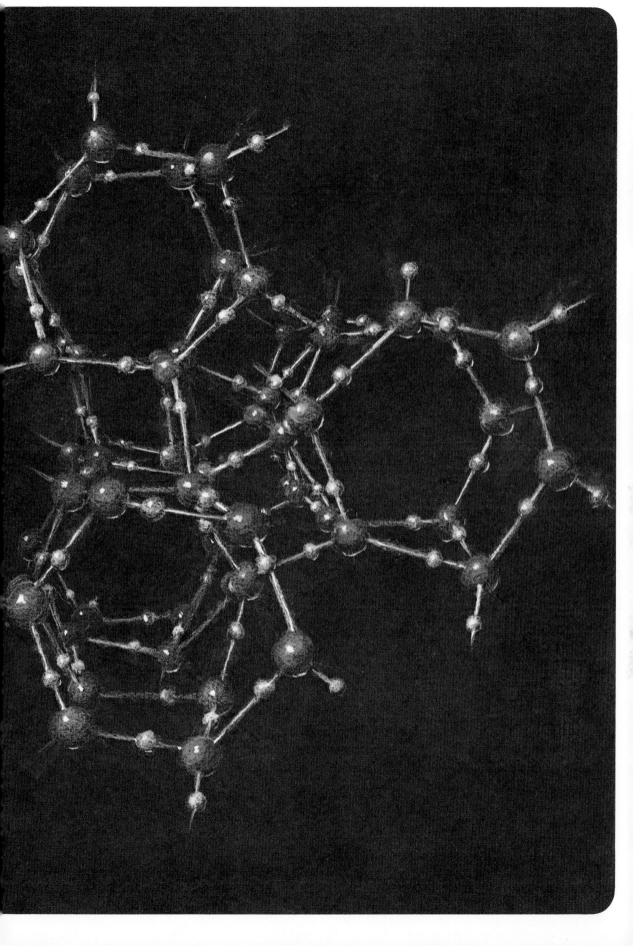

THE PENTAGONAL DODECAHEDRON

43

The pentagonal dodecahedron is the fifth regular polyhedron. It has thirty edges, twenty corners, and twelve faces, which are regular pentagons. It is closely related to the icosahedron: the relation involves interchanging corners and faces. Its symmetry elements are the same as those for the icosahedron.

The significance of the pentagonal dodecahedron to molecular architecture is in part the result of the close approximation of the angles between its edges (108°, the angle characteristic of the regular pentagon) to the tetrahedral angle (109.5°). The hydrocarbon molecule $C_{20}H_{20}$, with twenty carbon atoms at the corners of a pentagonal dodecahedron (edge 1.54 Å, the customary value for the carbon-carbon single bond) and with a hydrogen atom 1.10 Å out from each carbon atom, would involve very little bond-angle strain. However, the orientation around each of the carbon-carbon bonds is the unstable one (plate 17). This feature of the structure may explain why chemists have not yet succeeded in synthesizing this hydrocarbon.

A structure involving the pentagonal dodecahedron is described on the following page.

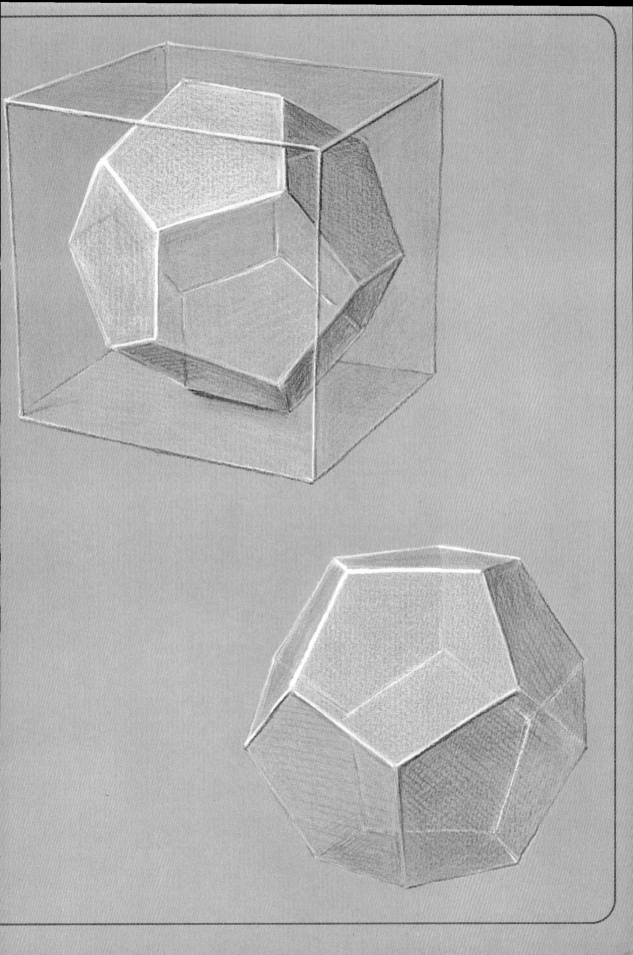

THE CLATHRATE CRYSTAL XENON HYDRATE

44

If twenty water molecules were placed at the corners of a pentagonal dodecahedron with edge 2.76 Å, they could use thirty of their hydrogen atoms to form unstrained hydrogen bonds along the dodecahedral edges. Several $(H_2O)_{20}$ aggregates of this sort can be seen in the adjacent drawing of a crystal with a framework of hydrogen-bonded water molecules.

The framework contains chambers of two kinds: the dodecahedral chambers of twenty water molecules, and somewhat larger chambers formed by twenty-four water molecules. Each of these larger chambers has two hexagonal faces and twelve pentagonal faces. A vertical column of them, with hexagonal faces shared, occupies the center foreground of the drawing.

This framework of water molecules is found in many crystals, such as xenon hydrate, $Xe \cdot 5\frac{3}{4}H_2O$ (or $8Xe \cdot 46H_2O$, the contents of the unit cube outlined in the drawing), and methane hydrate, $CH_4 \cdot 5\frac{3}{4}H_2O$. The drawing shows the xenon molecules (xenon atoms) in the chambers. Xenon hydrate can be classified with Prussian blue (plates 27 and 28) as a clathrate crystal.

This clathrate hydrate of xenon has special interest because of its relation to the theory of anesthesia. Xenon is an excellent anesthetic agent. Until recently no reasonable explanation of its anesthetic activity had been proposed. A recent suggestion is that xenon and other general anesthetic agents act by causing the water in the brain to form small clathrate crystals, which entrap ions and electrically charged groups of atoms and prevent them from contributing to the electric oscillations in the brain that constitute the mental activity characteristic of consciousness and sensitivity.

THE MOLECULE OF GLYCINE, THE SIMPLEST AMINO ACID

45

All living organisms contain molecules of proteins. Proteins are complicated substances, with a great many atoms in each molecule. During the past few years much information has been obtained about their molecular structure.

About a hundred years ago it was discovered that when any protein is boiled with acid and the resulting solution is allowed to stand, crystals of different substances separate from the solution. Glycine is one of the substances obtained in this way. It is called an amino acid, and it is the simplest of the amino acids obtained by the decomposition of proteins.

The structure of the glycine molecule as it exists in solution in the aqueous body fluids is shown in the adjacent drawing.

The formula of glycine was for many years written as

$$H_2N-CH_2-C\overset{\displaystyle O}{\underset{\displaystyle O-H}{\big<}}$$

In this formula there is shown an acidic group, COOH, identical with the acidic group in acetic acid (plate 40). This acidic group can liberate a proton into an aqueous solution. There is also shown the basic group NH_2, called the amino group. A substance with this group in its molecule is a base. It has the property of adding a proton, to produce a positive ion.

About fifty years ago chemists recognized that the glycine molecule in aqueous solution has the proton removed from the acidic group and attached to the nitrogen atom, as shown in the drawing. The acidic end of the molecule then carries a negative electric charge, and the basic end carries a positive electric charge. The molecule as a whole is electrically neutral.

In the drawing of the molecular structure, one oxygen atom is indicated to be bonded to the adjacent carbon atom by a double bond, and the other by a single bond. In fact, studies of the molecular structure have shown that the two carbon-oxygen distances are equal, with the value 1.25 Å. This fact is accounted for by saying that the double bond resonates between the two oxygen atoms.

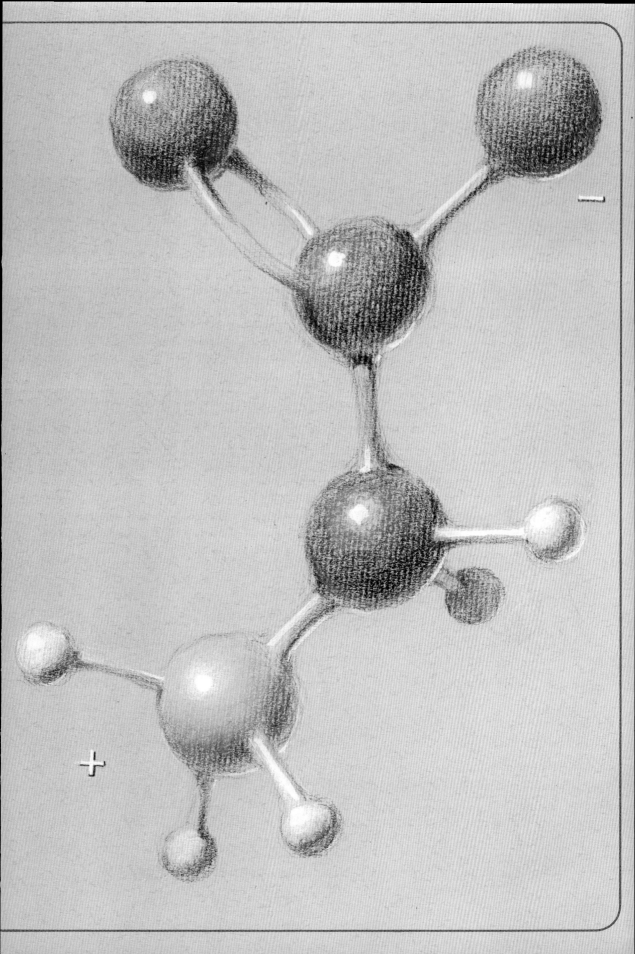

LEFT-HANDED AND RIGHT-HANDED MOLECULES OF ALANINE

46

Another amino acid that is obtained by the decomposition of proteins is L-alanine. It is closely similar to glycine in structure, but with one of the hydrogen atoms of glycine replaced by a methyl group, CH_3, as shown in the adjacent drawing.

There are two kinds of alanine molecules, which differ in the arrangement of the four groups around the central carbon atom. These molecules are mirror images of one another. The molecules of one kind are called D-alanine (D for Latin *dextro*, right), and those of the other kind L-alanine (L for Latin *laevo*, left). Only L-alanine occurs in living organisms as part of the structure of protein molecules.

Other amino acids, with the exception of glycine, also may exist both as D molecules and as L molecules, and in every case it is the L molecule that is involved in the protein molecules of living organisms. Some of the D amino acids cannot serve as nutrients, and may be harmful to life.

In *Through the Looking-Glass* Alice said, "Perhaps looking-glass milk isn't good to drink." When this book was written, in 1871, nobody knew that protein molecules are built of the left-handed amino acids; but Alice was justified in raising the question. The answer is that looking-glass milk is not good to drink.

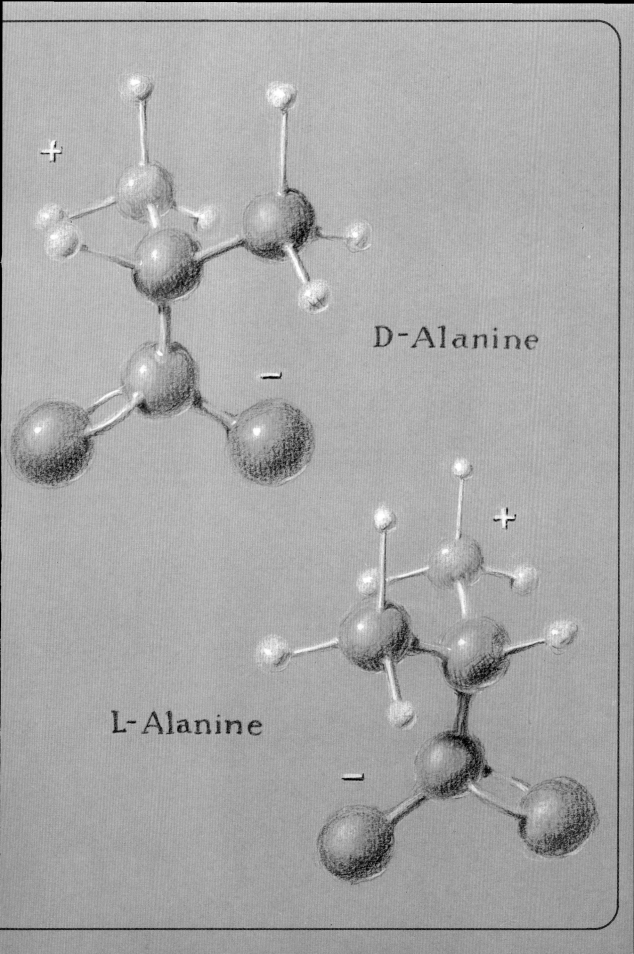

D-Alanine

L-Alanine

THE GLYCYLGLYCINE MOLECULE

47

The way in which amino-acid molecules combine to form proteins is illustrated by the adjacent drawing, which shows the structure of a molecule formed by the reaction of two molecules of the amino acid glycine with one another. A molecule of water is also produced in this reaction:

$$H_3\overset{+}{N}-CH_2-CO\overset{-}{O} + H_3\overset{+}{N}-CH_2-CO\overset{-}{O} \longrightarrow$$

$$H_3\overset{+}{N}-CH_2-CO-NH-CH_2-CO\overset{-}{O} + H_2O$$

The product glycylglycine is called a peptide. It is described as containing two glycine residues. A polypeptide contains many amino-acid residues in a long chain.

The characteristic structural feature of peptides is the six-atom group

It is called the peptide group. In this group of six atoms the double bond is shown from a carbon atom to the oxygen atom. In fact, the double bond may be placed between the carbon atom and the nitrogen atom, with a single bond to the oxygen atom; that is, the double bond may be described as resonating between the two positions. The double-bond character of the carbon-nitrogen bond requires that the six atoms lie in one plane. This feature of molecular architecture, the planarity of the peptide group, has been verified by the determination of the crystal structure of glycylglycine and other peptides. All of the dimensions of the peptide group have been determined with an accuracy of about 0.01 Å.

THE MOLECULAR ARCHITECTURE OF SILK

48

The fibers of silk that are spun by silkworms and spiders consist mainly of the protein called silk fibroin. Different species of silkworms and spiders produce different kinds of silk fibroin, which differ somewhat from one another in their molecular architecture. All of the kinds of silk fibroin contain long zigzag polypeptide chains lying parallel to one another, as shown in the adjacent drawing. The chains extend in the direction of the fiber or thread (vertically in the drawing).

The drawing shows a small portion of a single layer of protein molecules (polypeptide chains) in a silk fiber. These molecules are attached to adjacent molecules in the layer by the formation of hydrogen bonds, which extend laterally from the NH group of one chain to the oxygen atom of an adjacent chain. The fiber consists of many of these hydrogen-bonded layers superimposed upon one another.

The green spheres in the drawing are the side chains of the various amino acids whose residues comprise the protein molecules. In ordinary commercial silk, spun by the silkworm *Bombyx mori*, every other residue in each protein molecule is a residue of glycine; all of the green spheres on one side of the layer represent hydrogen atoms. Most of the other residues are residues of L-alanine; the green spheres on the other side of the layer represent the methyl group, CH_3. In wild silk, made by the silkworm *Antherea pernyi*, most of the side chains on both sides of the layer are methyl groups.

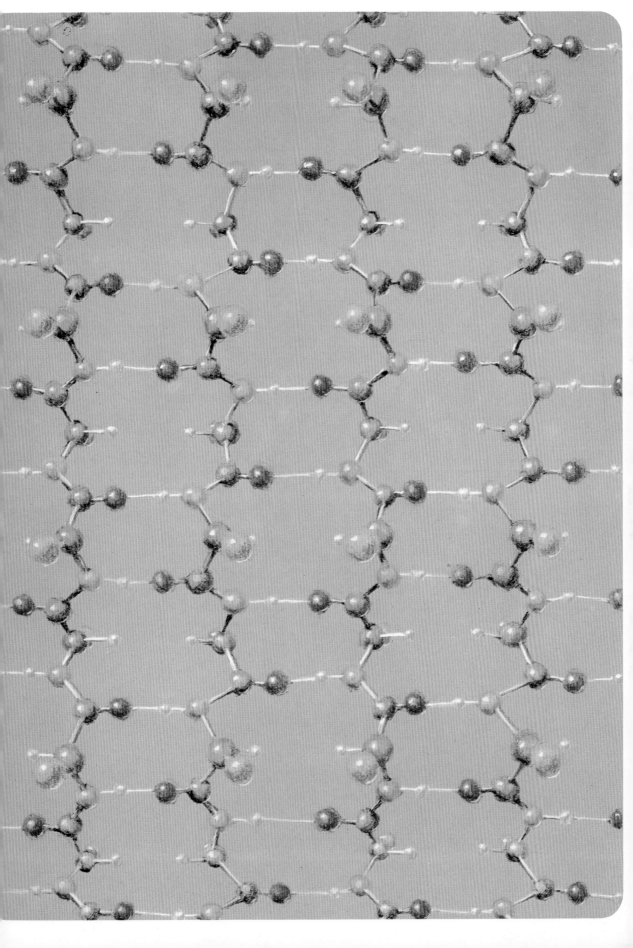

FOLDING THE POLYPEPTIDE CHAIN: A PROBLEM IN MOLECULAR ARCHITECTURE

49

In the period around 1920, x-ray diffraction patterns of silk, hair, muscle, tendon, and other fibrous proteins were first made, and the surmise (later verified) was advanced that in silk fibroin the protein molecules are in an extended configuration, as shown in the preceding drawing.

The British scientist William T. Astbury and his coworkers showed in 1931 that hair, muscle, horn, and porcupine quill give a characteristic x-ray pattern, and that when hair is steamed and stretched to double its original length the pattern changes to one resembling that of silk. They concluded that the polypeptide chains in unstretched hair are folded, and for nearly twenty years many scientists interested in molecular structure tried to determine the nature of the folding.

The problem was solved in 1950, when the alpha helix, shown in the adjacent drawing, was discovered through the application of knowledge obtained by the study of simple substances.

The peptide group is planar. There is considerable freedom of rotation about the single bonds to the carbon atom between two peptide groups. Stable ways of folding the polypeptide chain by rotating around these single bonds are those ways that permit the formation of N—H – – – O hydrogen bonds, with a distance of 2.79 Å between the nitrogen atom and the oxygen atom.

The alpha helix satisfies this requirement. There are 3.6 to 3.7 amino-acid residues per turn of the helix (about eighteen residues in five turns), and the nitrogen atom of each planar peptide is bonded to the oxygen atom of the third peptide beyond it in the chain by a hydrogen bond with length 2.79 Å.

MORE ABOUT THE ALPHA HELIX

50

A segment of an alpha helix is shown in the adjacent drawing. This segment contains twenty-three amino-acid residues and comprises about 6.5 turns of the helix. It is about 35 Å long; the length of the alpha helix is 1.49 Å per amino-acid residue.

In hair, horn, muscle, fingernail, and porcupine quill there are very long protein molecules with the alpha-helix structure extending in the fiber direction (along the length of a hair, sideways in a fingernail). They are not exactly parallel to one another, but are twisted around one another to form ropes of three or seven alpha helixes.

An alpha helix of a polypeptide might resemble either a right-handed screw or a left-handed screw. The segment shown in the drawing is a right-handed helix of L-amino-acid residues. This is the kind of alpha helix that has been found in proteins; no protein has been shown to contain left-handed alpha helixes.

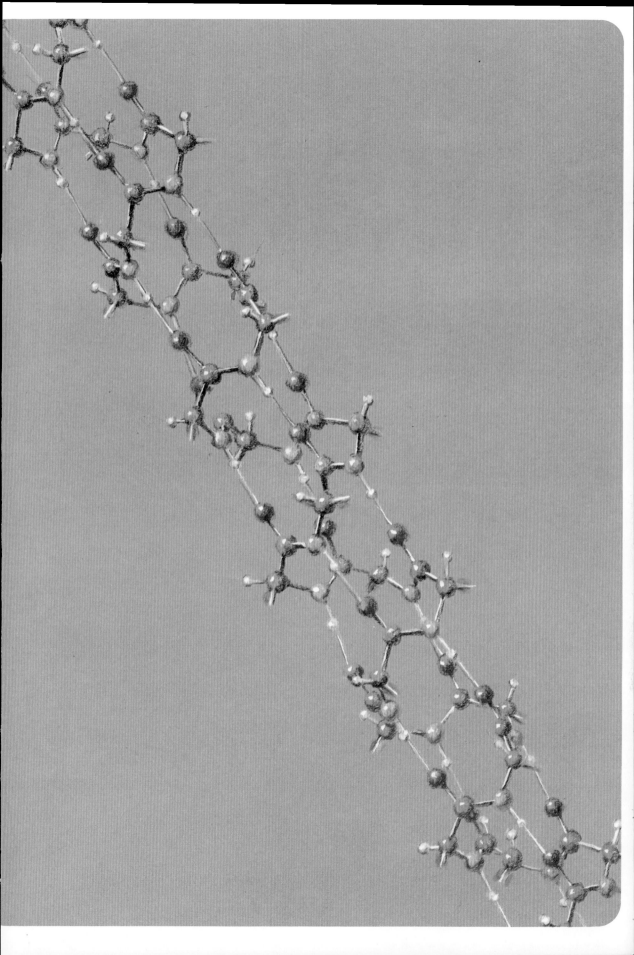

A PART OF THE MYOGLOBIN MOLECULE

51

Myoglobin is a substance found in muscle that resembles the substance hemoglobin that is found in the red cells of the blood. The molecule of myoglobin is made of about 2,500 atoms. The structure of myoglobin was determined, after many years of effort, by the British scientist John C. Kendrew (born 1917), by the x-ray investigation of myoglobin crystals. The British scientist Max F. Perutz (born 1914) made a similar study of hemoglobin, whose molecules are four times larger.

The structure of a portion of the myoglobin molecule, as determined by Kendrew, is shown in the adjacent drawing. The molecule contains 151 amino-acid residues, which constitute one polypeptide chain. The chain forms eight alpha-helix segments; one of them, showing five turns of the helix, is in the left foreground of the drawing.

The large atom just below and to the right of the center is the iron atom of myoglobin, to which an oxygen molecule attaches itself when myoglobin performs its function of storing oxygen that has been brought to the muscle by the hemoglobin in the blood. The iron atom is the central atom of a nearly planar group of seventy-three atoms called heme. Toward the lower left from the iron atom there is a ring of five atoms, part of a residue of the amino acid histidine. This residue and the heme molecule are discussed on the following pages.

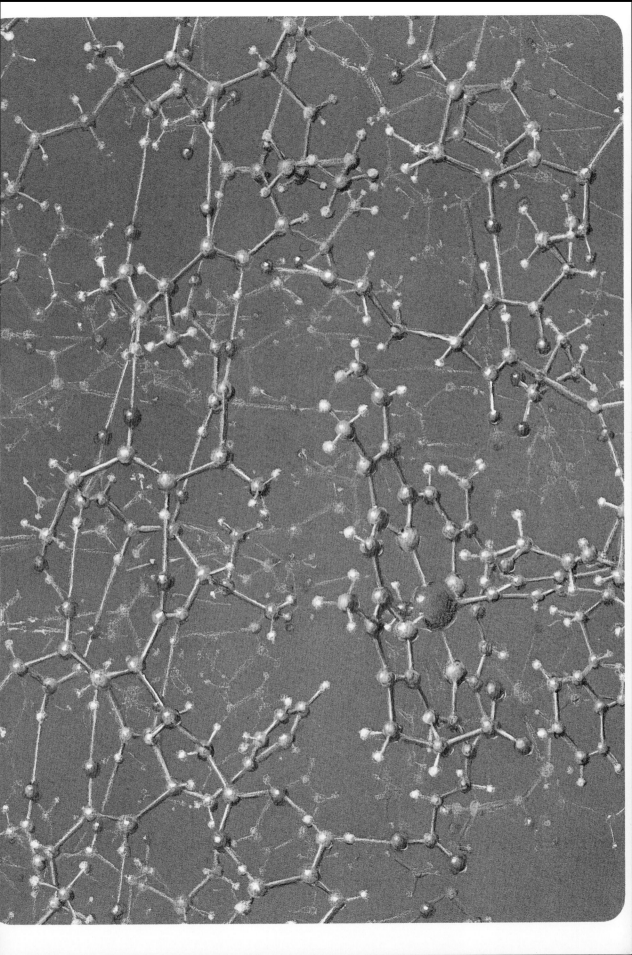

THE STRUCTURE OF THE HEME MOLECULE

52

Heme is a substance with the formula $FeC_{34}H_{32}O_4N_4$. The myoglobin molecule contains one heme and the hemoglobin molecule contains four hemes. In these molecules the two acidic side chains of the heme are ionized by loss of protons, giving the group two negative electric charges (on the oxygen atoms).

The heme molecule is nearly planar, as represented in the adjacent drawing. The property of planarity is ascribed to the many double bonds in the molecule. These double bonds are not restricted to the positions shown in the structural formula, but resonate to other positions also.

In the heme molecule the iron atom forms bonds with the four nitrogen atoms of the heme. In myoglobin and hemoglobin it is also bonded to a nitrogen atom of an amino-acid side chain, on one side of the heme plane, and, when oxygenated, to an oxygen atom of the oxygen molecule, on the other side of the plane. The iron atom then shows octahedral ligancy.

Heme is strongly colored. Heme with an attached oxygen molecule is responsible for the red color of oxygenated blood and muscle and without oxygen for the bluish color of deoxygenated blood and muscle.

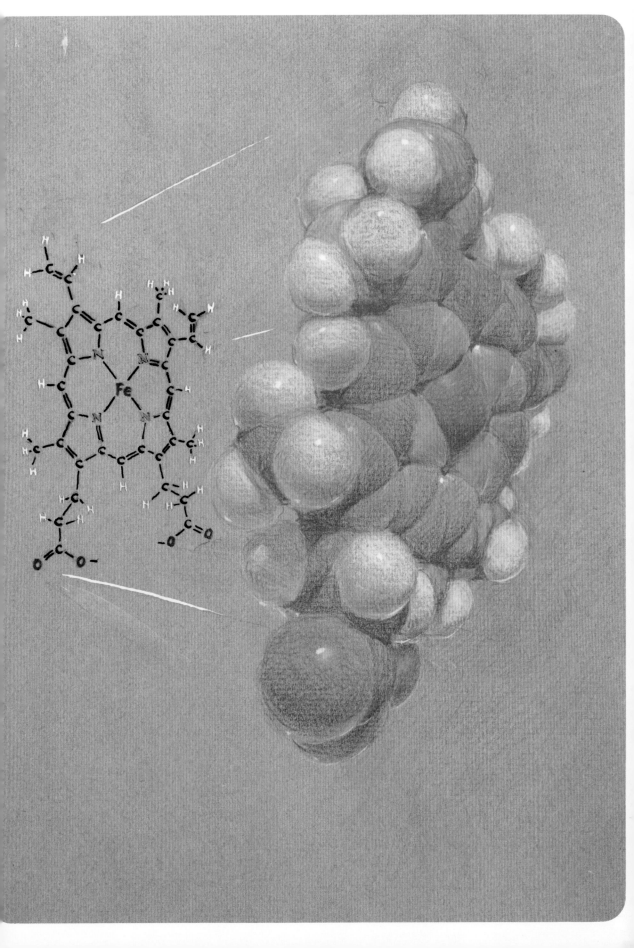

THE IRON ATOM IN
HEMOGLOBIN AND MYOGLOBIN

53

Iron forms many compounds. In some of them, such as ferrous oxide, FeO, the iron atom is said to be bivalent; in ferrous oxide it may be described as having transferred two electrons to the oxygen atom, leaving it a doubly charged ion, Fe^{++}. In other compounds, such as ferric oxide (the mineral hematite), Fe_2O_3, the iron atom has valence three, corresponding to the triply charged ion, Fe^{+++}. The ferric compounds are more stable than the ferrous compounds.

Hemoglobin and myoglobin normally contain ferrous iron atoms, Fe^{++}. Under certain conditions the iron atoms can be changed to the ferric state, Fe^{+++}. These ferric compounds, called ferrihemoglobin and ferrimyoglobin, do not have the power of combining reversibly with oxygen molecules.

An interesting problem is presented by the fact that in myoglobin and hemoglobin the ferrous state of iron atoms is more stable than in other iron compounds. The answer is connected with the presence of the histidine residue shown in the preceding drawing and again (from a different view) in the adjacent drawing.

The side chain of the amino acid histidine is $CH_2C_3N_2H_3$. The $C_3N_2H_3$ group is a five-membered ring. In blood and muscle this group acts as a basic group; it adds a proton, to become $C_3N_2H_4^+$, as shown in the drawing. Scientists have concluded that the presence of this positive electric charge on the side chain of the histidine residue that is in the sixty-second position in the polypeptide chain of myoglobin (and similar positions for hemoglobin) serves to hinder the change from Fe^{++} to Fe^{+++} by its electrostatic repulsion of the extra positive charge characteristic of the ferric state of the iron atom.

A MOLECULAR DISEASE

54

During recent years several diseases have been shown to be molecular diseases. They involve the manufacture by the patient of molecules that are abnormal; that is, that have a structure different from the molecules manufactured by other people.

Some people have a sort of anemia in which the blood carries only half the normal amount of oxygen from the lungs to the tissues, although there is the normal amount of hemoglobin in the blood.

The hemoglobin molecule contains four polypeptide chains and four heme groups. Two of the polypeptide chains, called the alpha chains, are chains of 141 amino-acid residues; the other two, called beta chains, are chains of 146 amino-acid residues. In some anemic patients there are abnormal alpha chains or abnormal beta chains, such that the iron atom associated with the abnormal chains is easily converted to the ferric state, losing its power to combine with an oxygen molecule. This sort of anemia is called ferrihemoglobinemia or methemoglobinemia.

For some patients with this disease the molecular abnormality (which is caused by a mutated gene) is the replacement of the histidine residue in the fifty-eighth position of the alpha chain or the sixty-third position of the beta chain by a tyrosine residue, shown in the adjacent drawing. The tyrosine side chain, $CH_2C_6H_4OH$, does not have the property of adding a proton and assuming the positive electric charge that would stabilize the ferrous state of the iron atom.

Knowledge of the molecular structure of normal and abnormal hemoglobin thus provides an essentially complete explanation of the manifestations of this disease.

A MOLECULAR ABNORMALITY
THAT DOES NOT CAUSE
A DISEASE

55

Some people have been found to have an abnormality in the sixty-third position in the beta chain of their hemoglobin molecules, but to be free of the disease ferrihemoglobinemia.

Their gene mutation causes the histidine residue in this position to be replaced by an arginine residue. The side chain of arginine is

This is a basic group, which adds a proton to form the positively charged group shown in the adjacent drawing.

The positive electric charge of the arginine side chain serves the same function as the charge of the histidine side chain that is normally present, stabilizing the ferrous state of the iron atom. We can accordingly understand why the molecular abnormality involving tyrosine in this position leads to the disease ferrihemoglobinemia and the abnormality involving arginine does not. The abnormality involving arginine does, however, cause a sensitivity of the red cells to certain drugs, the sulfonamides, which then leads to severe anemia. The mechanism of this effect is not known.

Molecular diseases were discovered only fifteen years ago. Much has been learned since, but the possibilities for further discovery are tremendous. Knowledge that may be obtained in the near future about the molecular structure of the human body and the molecular basis of disease may well lead to a great decrease in the amount of human suffering.

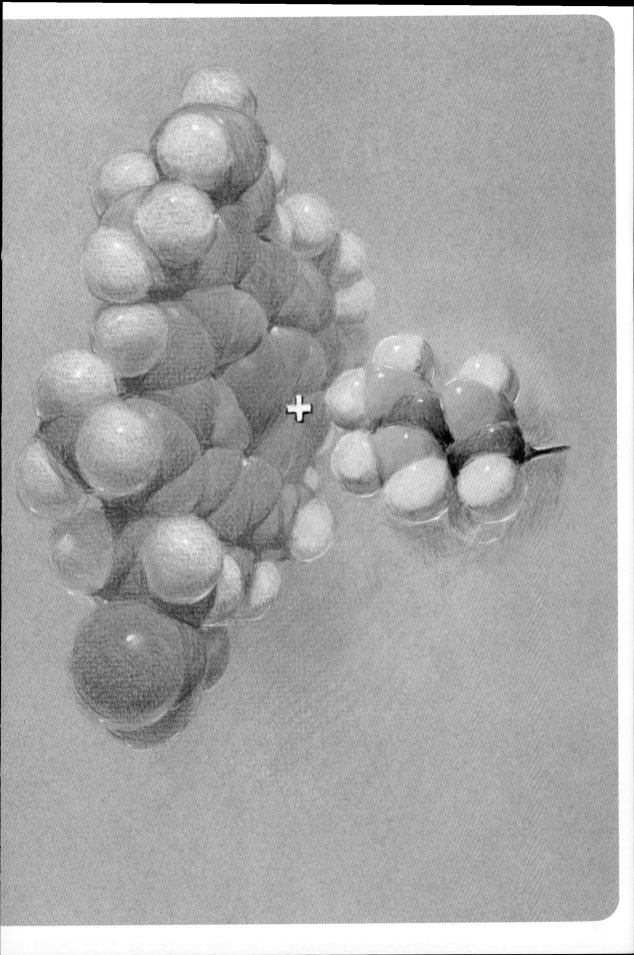

MOLECULAR COMPETITION: SULFANILAMIDE AND PARA-AMINOBENZOIC ACID

56

Thirty years ago the infectious diseases constituted the principal cause of death. Now most of these diseases have been brought under control. The recent period of rapid progress began with the discovery of the sulfa drugs in 1935 by the German chemist Gerhard Domagk (1895–1964). The structure of the molecule of sulfanilamide, one of the sulfa drugs, is shown at the upper left in the adjacent drawing.

Although the molecular structures of many drugs are known, we are still for the most part ignorant about the mechanism of action of the drugs. Sulfanilamide is an exception. There is good evidence that the sulfanilamide molecule acts by entering into competition with molecules of para-aminobenzoic acid (bottom right of the drawing). The bacteria in a culture containing a little para-aminobenzoic acid cease to grow when sulfanilamide is added to the culture; they resume growth if more para-aminobenzoic acid is added; they again cease to grow if more sulfanilamide is added.

Para-aminobenzoic acid seems to be a bacterial growth vitamin. It is likely that a para-aminobenzoic acid molecule fits into a cavity of a bacterial protein molecule, and then carries out some function essential to growth. The sulfanilamide molecule ($H_2NC_6H_4SO_2NH_2$) closely resembles the para-aminobenzoic acid molecule ($H_2NC_6H_4CO_2H$) in size and shape, as can be seen from the drawing. It is accordingly reasonable to postulate that if enough sulfanilamide molecules are present in the bacteria, they may, in accordance with the principles of chemical equilibrium, occupy the cavities in the protein molecules and prevent the molecules of the bacterial vitamin from entering these cavities and carrying out their growth-promoting function.

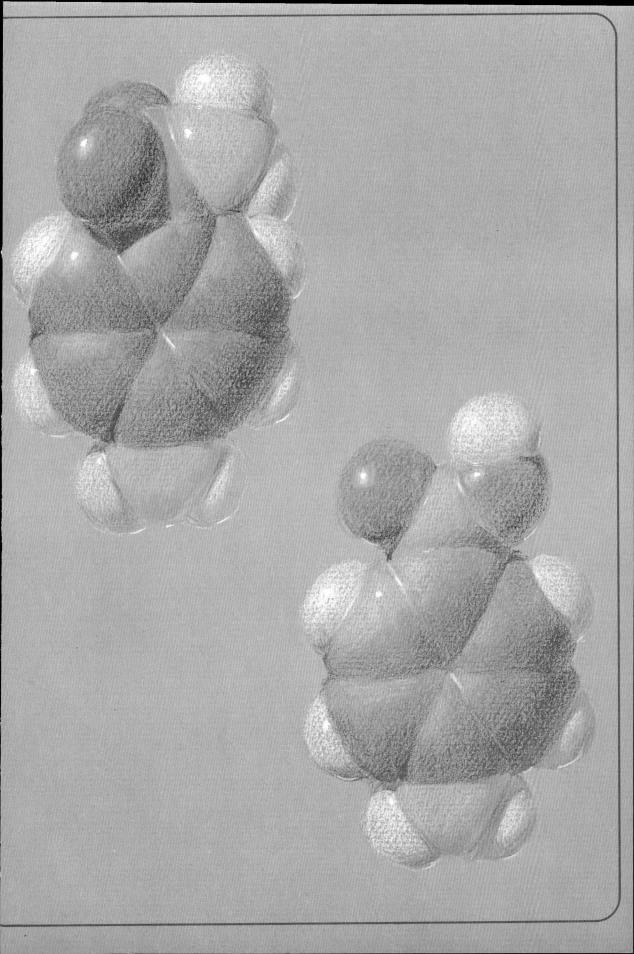

AN ANTIVIRAL MOLECULE

57

The sulfa drugs and many other substances, such as penicillin, are effective against bacteria but not against viruses. Some substances with the power of controlling certain viral infections have been discovered in recent years. One of these substances is chlortetracycline, $C_{22}H_{23}O_8N_2Cl$.

Chlortetracycline (also called Aureomycin) is a golden-yellow substance that is made by the mold *Streptomyces aureofaciens*. The structure of the chlortetracycline molecule has been precisely determined by the x-ray diffraction method; it is shown in the adjacent drawing. One of the features of the structure is the large number of hydrogen bonds formed between adjacent oxygen atoms in the molecule. Another is the sequence of four six-membered rings, fused together (referred to in its name). The molecule contains seven double bonds (some resonating), which are not indicated in the drawing.

Little is known as yet about the molecular mechanisms by means of which chlortetracycline is able to control some viral diseases and penicillin (the formula of penicillin G is $C_{16}H_{18}O_4N_2S$) is able to control many bacterial diseases. Knowledge of the molecular structure of chlortetracycline and penicillin does not by itself give the solution of the great problem of the molecular basis of the action of drugs and the nature of disease. We need also to know the molecular structure of the human body, of bacteria, of viruses. When these problems have been solved it will be possible to apply much of our present knowledge, as well as the new knowledge, in a way that will benefit all humanity.

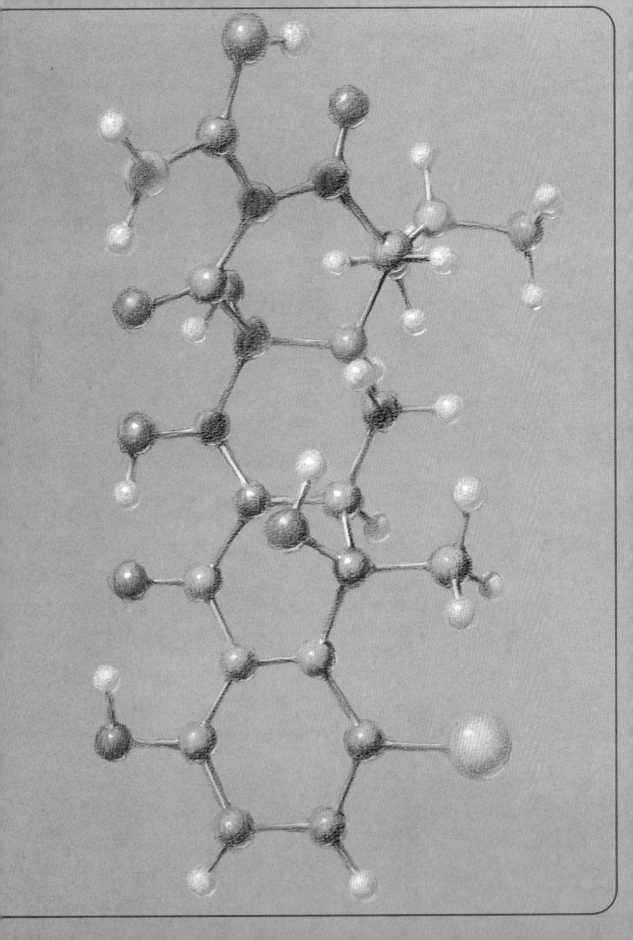

THE PERIODIC SYSTEM OF THE ELEMENTS

H 1	He 2	

GROUP →

	I	II	III	IV	V	VI	VII	
He 2	Li 3	Be 4	B 5	C 6	N 7	O 8	F 9	Ne 10
Ne 10	Na 11	Mg 12	Al 13	Si 14	P 15	S 16	Cl 17	Ar 18

GROUP →

	I	II	III	IVa	Va	VIa	VIIa	VIII			Ib	IIb	IIIb	IV	V	VI	VII	
Ar 18	K 19	Ca 20	Sc 21	Ti 22	V 23	Cr 24	Mn 25	Fe 26	Co 27	Ni 28	Cu 29	Zn 30	Ga 31	Ge 32	As 33	Se 34	Br 35	Kr 36
Kr 36	Rb 37	Sr 38	Y 39	Zr 40	Nb 41	Mo 42	Tc 43	Ru 44	Rh 45	Pd 46	Ag 47	Cd 48	In 49	Sn 50	Sb 51	Te 52	I 53	Xe 54
Xe 54	Cs 55	Ba 56	La 57 *	Hf 72	Ta 73	W 74	Re 75	Os 76	Ir 77	Pt 78	Au 79	Hg 80	Tl 81	Pb 82	Bi 83	Po 84	At 85	Rn 86
Rn 86	Fr 87	Ra 88	Ac 89 ♦															

*Lanthanons

Ce 58	Pr 59	Nd 60	Pm 61	Sm 62	Eu 63	Gd 64	Tb 65	Dy 66	Ho 67	Er 68	Tm 69	Yb 70	Lu 71

♦Actinons

Th 90	Pa 91	U 92	Np 93	Pu 94	Am 95	Cm 96	Bk 97	Cf 98	Es 99	Fm 100	Md 101	No 102	Lw 103

VALUES OF PACKING RADII OF ATOMS

Packing radii give the effective size for nonbonded contacts between atoms packed together in a crystal or liquid. They are also called Van der Waals radii, after the Dutch physicist J. D. van der Waals (1837–1923).

H	1.15 Å	N	1.5Å	O	1.40 Å	F	1.36Å
		P	1.9	S	1.85	Cl	1.81
		As	2.0	Se	2.00	Br	1.95
		Sb	2.2	Te	2.20	I	2.16

ATOMIC NUMBER	SYMBOL	ELEMENT	ATOMIC NUMBER	SYMBOL	ELEMENT	ATOMIC NUMBER	SYMBOL	ELEMENT
		Hydrogen	36	Kr	Krypton	70	Yb	Ytterbium
		Helium	37	Rb	Rubidium	71	Lu	Lutetium
		Lithium	38	Sr	Strontium	72	Hf	Hafnium
		Beryllium	39	Y	Yttrium	73	Ta	Tantalum
		Boron	40	Zr	Zirconium	74	W	Tungsten
		Carbon	41	Nb	Niobium	75	Re	Rhenium
		ogen	42	Mo	Molybdenum	76	Os	Osmium
		en	43	Tc	Technetium	77	Ir	Iridium
		e	44	Ru	Ruthenium	78	Pt	Platinum
			45	Rh	Rhodium	79	Au	Gold
		um	46	Pd	Palladium	80	Hg	Mercury
		m	47	Ag	Silver	81	Tl	Thallium
			48	Cd	Cadmium	82	Pb	Lead
		s	49	In	Indium	83	Bi	Bismuth
			50	Sn	Tin	84	Po	Polonium
			51	Sb	Antimony	85	At	Astatine
			52	Te	Tellurium	86	Rn	Radon
			53	I	Iodine	87	Fr	Francium
			54	Xe	Xenon	88	Ra	Radium
		m	55	Cs	Cesium	89	Ac	Actinium
			56	Ba	Barium	90	Th	Thorium
			57	La	Lanthanum	91	Pa	Protactinium
23		ium	58	Ce	Cerium	92	U	Uranium
24		omium	59	Pr	Praseodymium	93	Np	Neptunium
25		nganese	60	Nd	Neodymium	94	Pu	Plutonium
26		ron	61	Pm	Promethium	95	Am	Americium
27	Co	Cobalt	62	Sm	Samarium	96	Cm	Curium
28	Ni	Nickel	63	Eu	Europium	97	Bk	Berkelium
29	Cu	Copper	64	Gd	Gadolinium	98	Cf	Californium
30	Zn	Zinc	65	Tb	Terbium	99	Es	Einsteinium
31	Ga	Gallium	66	Dy	Dysprosium	100	Fm	Fermium
32	Ge	Germanium	67	Ho	Holmium	101	Mv	Mendelevium
33	As	Arsenic	68	Er	Erbium	102	No	Nobelium
34	Se	Selenium	69	Tm	Thulium	103	Lw	Lawrencium
35	Br	Bromine						

VALUES OF SINGLE-BOND COVALENT RADII OF ATOMS

The interatomic distance for two atoms connected by a single bond is approximately equal to the sum of their covalent radii. The distance for a double bond is about 0.21 Å less and for a triple bond about 0.34 Å less than for a single bond.

H	0.30 Å	C	0.77 Å	N	0.70 Å	O	0.66 Å	F	0.64 Å
(0.37 Å in H_2)		Si	1.17	P	1.10	S	1.04	Cl	0.99
		Ge	1.22	As	1.21	Se	1.17	Br	1.14
		Sn	1.40	Sb	1.41	Te	1.37	I	1.33